Survival Guide for
General, Organic, and Biochemistry
With Math Review

exam 1 wed feb 20

Richard Morrison
University of Georgia

Charles H. Atwood
University of Georgia

Joel A. Caughran
University of Georgia

THOMSON

BROOKS/COLE

Australia • Brazil • Canada • Mexico • Singapore • Spain • United Kingdom • United States

Printer: Thomson/West
Cover Image: © Images.com/Corbis

ISBN-13: 978-0-495-55469-1
ISBN-10: 0-495-55469-3

Thomson Higher Education
10 Davis Drive
Belmont, CA 94002-3098
USA

For more information about our products, contact us at:
Thomson Learning Academic Resource Center
1-800-423-0563

For permission to use material from this text or product, submit a request online at
http://www.thomsonrights.com.
Any additional questions about permissions can be submitted by email to **thomsonrights@thomson.com.**

to Margaret, Laura, Daniel, William, and Emily
~ RWM

to Judy, Louis, and Lesley
~ CHA

to Tashia
~ JAC

Table of Contents

Module 4
Chemical Reactions and Stoichiometry

Module 5
States of Matter, Solutions and Colligative Properties

Module 6
Chemical Kinetics and Equilibrium

Module 7
Acids and Bases

Module 8
Functional Groups and Organic Nomenclature

Module 9
Stereoisomerism

Module 10
Substitution Reactions

Module 11
Addition Reactions

Module 12
Elimination Reactions

Module 13
Reduction-Oxidation Reactions

Module 14
Carbohydrates

Module 15
Lipids

Module 16
Proteins and Enzymes

Module 17
Nucleotides and Nucleic Acids

Math Review

Preface

The authors have written this GOB Survival Guide to accompany a conventional two-semester or three-quarter General, Organic and Biochemistry course. It is intended to complement rather than replace the course textbook. Topics are selected to provide further explanation and practice in areas where, experience has shown, additional discussion and problem-solving help is often necessary.

The GOB Survival Guide is partitioned into modules presenting conceptually related material which may be discussed in two or more chapters of a traditional textbook. For example, organic nomenclature is grouped together into one module, whereas traditional textbooks present nomenclature throughout the chapters describing each organic functional group. The modular grouping allows students to learn the common elements of nomenclature as they uniformly apply to all families of organic compounds and consolidates nomenclature into one readily accessed source. Similar conceptual groupings occur in the modular presentations of organic substitution, addition, and elimination reactions.

Every module begins with an introduction followed by boxed descriptions of **key concepts**. <u>**Sample exercises**</u> containing **tips** and **cautions** are included in every module to provide additional practice and rehearse problem solving skills. The general chemistry modules contain relatively brief descriptions of key concepts followed by sample exercises. Many of the organic and biochemistry modules contain expanded descriptions of key concepts intermingled with sample exercises. This minor variation in format is viewed by the authors as a helpful modification designed to address the more visually and structurally based concepts of organic chemistry.

The authors wish to thank our Thomson Learning colleagues David Harris, Lisa Lockwood, and Lisa Weber for their unwavering support and sound advice. We also are indebted to Dr. Mark Erickson for his insightful reviews of this manuscript, to our editorial assistants Sylvia Krick and Liz Woods, and to Dr. Angela Sauers and Dr. Richard Hubbard for their helpful suggestions. Finally, we are grateful to Margaret Morrison, Judy Atwood, and Tashia Caughran, our patient wives who have encouraged us throughout the writing and completion of this GOB Survival Guide.

Module 1
Metric System, Significant Figures,
Dimensional Analysis, Density and Heat Transfer

Introduction

This module addresses several topics that are typically introduced in the first chapter of general chemistry textbooks. This module describes:

1. how to use scientific notation
2. the basic rules of the metric system and significant figures
3. how to use dimensional analysis to solve problems
4. the relationship between density, mass, and volume and how to apply dimensional analysis to density problems
5. how heat is transferred from one object to another

Module 1 Key Concepts

1. $\text{density} = \dfrac{\text{mass}}{\text{volume}} = \dfrac{m}{v}$

2. $\text{amount of heat} = \text{specific heat} \times \text{mass} \times \text{change in temperature}$

    or

    $q = SH \times m \times \Delta T$

    For both equations, if any one of the variables in the equation is unknown, then you can solve for it using basic algebra.

The metric system uses a series of multipliers to convert from one sized unit to another size. You must be very familiar with these prefixes and how to convert from one size unit to another. A common set of multiplier prefixes is given in the table below.

Scientific Notation

In the physical and biological sciences it is frequently necessary to write numbers that are extremely large or small. It is not unusual for these numbers to have 20 or more digits beyond the decimal point. For the sake of simplicity and to save space when writing these numbers, a compact or shorthand method of writing, called scientific notation, is employed. In both methods the insignificant digits that are placeholders between the decimal place and the significant figures are expressed as powers of ten. Significant digits are then multiplied by the appropriate powers of ten to give a number that is both mathematically correct and indicative of the correct number of significant figures to use in the problem. To be strictly correct, the significant figures should be between 1.000

and 9.999; however, this particular rule is frequently ignored. In fact, it must be ignored when adding numbers in scientific notation that have different powers of ten.

TIP

> ***Positive powers of ten*** indicate that the decimal place has been ***moved to the left that number of spaces***.
>
> ***Negative powers of ten*** indicate that the decimal place has been ***moved to the right that number of spaces***.

A few examples of numbers written in scientific notation are given in this table.

Quiz 1

Number	Scientific Notation
10,000	1×10^4
100	1×10^2
1	1×10^0
0.01	1×10^{-2}
0.000001	1×10^{-6}
23,560	2.356×10^4
0.0000965	9.65×10^{-5}

You must understand how to use scientific notation to express very large or small numbers. Familiarize yourself with this method.

Metric System
It is important to recognize that the metric prefixes may be used with any unit of measurement, and that the relationship between the *base unit* and the unit with the prefix is always the same regardless of the base unit. The base unit is represented by x in the table.

Pay special attention to the unit factors provided as they are what will be used in converting one unit to another. Note that each unit factor may be written in two equivalent ways. The one you use depends on what units you are trying to cancel in a dimensional analysis problem (see examples below).

One way to help insure that you work conversion problems correctly is to remember which one of the units is the largest. For example, if you are converting from pg to Mg, then keep in mind that a Mg is much, much larger than a pg. So, the numerical value should get much smaller as you convert from pg to Mg.

Quiz 1

Prefix Name	Prefix Symbol	Multiplication Factor	Unit Factors
mega-	M	1000000 or 10^6	$\dfrac{1\,Mx}{10^6\,x} = \dfrac{10^6\,x}{1\,Mx}$
kilo-	k	1000 or 10^3	$\dfrac{1\,kx}{10^3\,x} = \dfrac{10^3\,x}{1\,kx}$
deci-	d	0.1 or 10^{-1} $\frac{1}{10}$	$\dfrac{1\,dx}{10^{-1}\,x} = \dfrac{10^{-1}\,x}{1\,dx}$
centi-	c	0.01 or 10^{-2} $\frac{1}{100}$	$\dfrac{1\,cx}{10^{-2}\,x} = \dfrac{10^{-2}\,x}{1\,cx}$
milli-	m	0.001 or 10^{-3} $\frac{1}{1000}$	$\dfrac{1\,mx}{10^{-3}\,x} = \dfrac{10^{-3}\,x}{1\,mx}$
micro-	μ	0.000001 or 10^{-6} $\frac{1}{1\,000\,000}$	$\dfrac{1\,\mu x}{10^{-6}\,x} = \dfrac{10^{-6}\,x}{1\,\mu x}$
nano-	n	0.000000001 or 10^{-9} $\frac{1}{100\,000\,000}$	$\dfrac{1\,nx}{10^{-9}\,x} = \dfrac{10^{-9}\,x}{1\,nx}$
pico-	p	0.000000000001 or 10^{-12} $\frac{1}{1\,000\,000\,000\,000}$	$\dfrac{1\,px}{10^{-12}\,x} = \dfrac{10^{-12}\,x}{1\,px}$

Sample Exercises

1. How many mm are there in 3.45 km?
   The correct answer is 3.45×10^6 mm

The table indicates that there are 1000 m in 1 km and that 1 mm = 0.001 m.

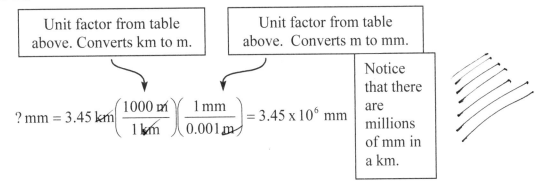

| Unit factor from table above. Converts km to m. | Unit factor from table above. Converts m to mm. | Notice that there are millions of mm in a km. |

$$? \, mm = 3.45 \, \cancel{km}\left(\frac{1000 \, \cancel{m}}{1 \, \cancel{km}}\right)\left(\frac{1 \, mm}{0.001 \, \cancel{m}}\right) = 3.45 \times 10^6 \, mm$$

Note that the km and the m both cancel. The canceling of units is the key to dimensional analysis problems.

In this problem, the km is a much larger unit than the mm. Thus we should expect that there will be many of the smaller unit, mm's, in the large units. The answer 3.45×10^6 mm is sensible.

TIP Always convert to a base unit (like m or g) first. Then proceed to a different unit if necessary.

2. How many mg are there in 15.0 pg?
The correct answer is 1.5 x 10⁻⁸ mg

From the table we see that 1 pg = 10^{-12} g and 1 mg = 10^{-3} g.

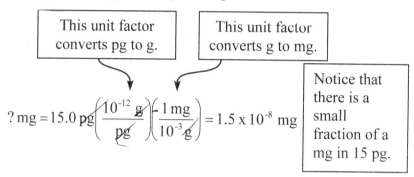

This unit factor converts pg to g.

This unit factor converts g to mg.

Notice that there is a small fraction of a mg in 15 pg.

$$? \, mg = 15.0 \, pg \left(\frac{10^{-12} \, g}{pg}\right)\left(\frac{1 \, mg}{10^{-3} \, g}\right) = 1.5 \times 10^{-8} \, mg$$

In this problem picograms, pg, are the smaller unit. We should expect that there are very few milligrams, mg, in 15.0 pg. The correct answer is 1.5×10^{-8} mg, which is reasonable.

Significant Figures   Quiz I Wed

All non-zero integers are significant. When determining the number of significant figures in a value, by far the most confusion revolves around zeros. Sometimes they are significant, and sometimes they are not! Below are some rules to help you determine whether or not a zero is significant.

1. Zeros located between two integers ARE significant.
2. Zeros located at the ends of numbers containing decimals ARE significant.
3. Zeros located between an integer on the right and a decimal on the left ARE significant.
4. Zeros used as place holders to indicate the position of a decimal ARE NOT significant. This includes a zero at the end of a number that does not contain a decimal.

<u>**Sample Exercises**</u>
3. How many significant figures are in the number 58062?
The correct answer is: 5 significant figures

This zero is significant because it is embedded in other significant digits. See rule 1.

4. How many significant figures are in the number 0.0000543?
The correct answer is: 3 significant figures

None of these zeroes are significant because their purpose is to indicate the position of the decimal place. Only the non-zero integers are significant.

5. How many significant figures are in the number 0.009120?
   The correct answer is: 4 significant figures

> This zero is significant.
> See rule 4!

> These three zeroes are not significant because they are place holders.

6. How many significant figures are in the number 24500?
   The correct answer is: 3 significant figures

> These zeroes are not significant since there is no decimal at the end of the number.

7. How many significant figures are in the number 2.4500×10^4?
   The correct answer is: 5 significant figures

> As written both of these zeroes are significant because the number contains a decimal.

Notice that this number is the same as a previous exercise, but written in scientific notation. All of the same rules apply.

CAUTION

> None of the numbers in the 10^x portion of numbers written in scientific notation are significant.

Calculations and Significant Figures

Rules for determining the number of significant figures in the answer to a calculation depend on the mathematical operation being performed.

- In addition and subtraction problems, the final answer must contain no digits beyond the most doubtful digit in the numbers being added or subtracted.
- In multiplication and division problems involving significant figures the final answer must contain the same number of significant figures as the number with the least number of significant figures.

Sample Exercises

8. What is the sum of 12.674 + 5.3150 + 486.9?
   The correct answer is: 504.9

> This 9 is in the tenths decimal place. It is the most doubtful digit in the sum.

The most doubtful digit in each of the numbers is underlined 12.674, 5.3150, 486.9. Notice that the 486.9 has the most doubtful digit because the 9 is only in the tenths position and the other numbers are doubtful in the thousandths and ten thousandths positions. *The final answer must have the final digit in the tenths position.*

5

9. *What is the correct answer to this problem:* $2.6138 \times 10^6 - 7.95 \times 10^{-3}$?
   The correct answer is: 2.6138×10^6

> This 8 is the most doubtful digit in the sum. It is in the hundreds position.

The number 2.6138×10^6 can be also written as 2,613,800. Its most doubtful digit, the 8, is in the hundreds position. The other number, 7.95×10^{-3}, can be written as 0.00795. Its most doubtful digit, the 5, is in the one millionths position. Consequently, the final answer cannot extend beyond the 8 in 2.6138×10^6.

 TIP | When adding and subtracting, both numbers must be expressed to the same power of 10 to determine the most doubtful digit.

10. *What is the correct answer to this problem:* 47.893×2.64?
   The correct answer is: 126

> This number contains only 3 significant digits, so the answer can have only 3 significant figures.

11. *What is the correct answer to this problem:* $1.95 \times 10^5 \div 7.643 \times 10^{-4}$?
   The correct answer is 2.55×10^8

> This number contains 3 significant figures.

> This number contains 4 significant figures.

Just as in exercise 10, the number with fewest significant digits determines that the final answer must also have three significant digits.

Dimensional Analysis *Quiz 3 Mon*
In chemistry we often perform calculations that require changing from one set of units to a second set of units. Dimensional analysis is a convenient method to help convert units without making arithmetic errors. In this method, common conversion factors given in your textbook are arranged so that one set of units cancels, converting the problem to the second set of units.

Sample Exercises
12. *How many Mm are in 653 ft?*
   The correct answer is: 1.99×10^{-4} Mm.

$653 ft \quad \dfrac{12 in}{1 ft} \quad \dfrac{2.54 cm}{1 in}$

6

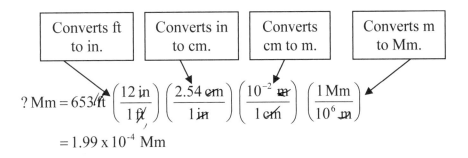

$$? \text{Mm} = 653 \text{ ft} \left(\frac{12 \text{ in}}{1 \text{ ft}} \right) \left(\frac{2.54 \text{ cm}}{1 \text{ in}} \right) \left(\frac{10^{-2} \text{ m}}{1 \text{ cm}} \right) \left(\frac{1 \text{ Mm}}{10^6 \text{ m}} \right)$$

$$= 1.99 \times 10^{-4} \text{ Mm}$$

Notice that the problem is arranged so that each successive conversion factor makes progress in the conversion process. Feet are converted to inches, then to cm, next to m, and finally to Mm. This is the simplest kind of dimensional analysis problem because all of the units are linear.

13. How many km² are in 2.5 x 10⁸ in²?
The correct answer is: 1.6 x 10⁻¹ km²

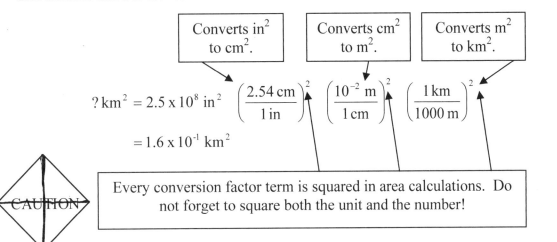

$$? \text{ km}^2 = 2.5 \times 10^8 \text{ in}^2 \left(\frac{2.54 \text{ cm}}{1 \text{ in}} \right)^2 \left(\frac{10^{-2} \text{ m}}{1 \text{ cm}} \right)^2 \left(\frac{1 \text{ km}}{1000 \text{ m}} \right)^2$$

$$= 1.6 \times 10^{-1} \text{ km}^2$$

CAUTION — Every conversion factor term is squared in area calculations. Do not forget to square both the unit and the number!

Because the problem involves area (a two dimensional unit) all of the conversion factors are similar to exercise 12, but they must be squared to be in the appropriate units.

TIP — The ( )² notation around a unit factor literally means that you are multiplying the unit factor by itself. If you have trouble remembering to square the unit factor, then try writing it out as the unit factor multiplied by the unit factor.

14. How many yd³ are in 7.93 x 10¹² cm³?
The correct answer is: 1.04 x 10⁷ yd³

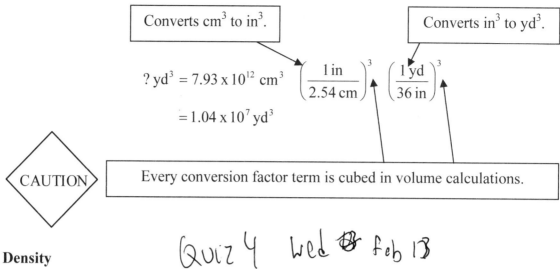

Converts cm³ to in³.		Converts in³ to yd³.

$$? \, yd^3 = 7.93 \times 10^{12} \, cm^3 \left(\frac{1 \, in}{2.54 \, cm}\right)^3 \left(\frac{1 \, yd}{36 \, in}\right)^3$$

$$= 1.04 \times 10^7 \, yd^3$$

CAUTION — Every conversion factor term is cubed in volume calculations.

Density

Quiz 4   Wed Feb 13

The ratio of an object's mass to its volume is the density. Many substances have a unique density which can be used to identify the substance. A closely related quantity is the specific gravity which is an object's density divided by the density of water. Since water's density is nearly 1.00 at room temperature, the numerical values for density and specific gravity are almost the same. However, specific gravity has no units.

Sample Exercises

15. **What is the mass, in g, of a 68.2 cm³ sample of ethyl alcohol? The density of ethyl alcohol is 0.789 g/cm³.**

    The correct answer is: 53.8 g

    Density converts the volume of a substance into the mass.

    $$D = \frac{m}{V} \Rightarrow m = DV$$

    $$? \, g = 68.2 \, cm^3 \left(\frac{0.789 \, g}{1 \, cm^3}\right)$$

    $$= 53.8 \, g$$

    The final units are g because the cm³ in the density cancels with the original volume.

16. **What is the volume, in cm³, of a 237.0 g sample of copper? The density of copper is 8.92 g/cm³.**

    The correct answer is: 26.6. cm³

    $$D = \frac{m}{V} \Rightarrow V = \frac{m}{D}$$

8

$$? \, cm^3 = 237.0 \, g\left(\frac{1 \, cm^3}{8.92 \, g}\right)$$

$$= 26.6 \, cm^3$$

17. **What is the density of a substance having a mass of 25.6 g and a volume of 74.3 cm³?**
    The correct answer is: 0.345 g/cm³

$$D = \frac{m}{V}$$

$$? \, g/cm^3 = \frac{25.6 \, g}{74.3 \, cm^3} = 0.345 \, g/cm^3$$

Density's units, g/cm³, help determine the correct order of division.

Quiz 4 wed feb 13

Heat Transfer

The amount of heat liberated or absorbed by a substance can be readily calculated in the following fashion. Specific heats of substances are experimentally measured and tabulated in your textbook.

Sample Exercise

18. **How much heat is required to heat 75.0 g of aluminum, Al, from 25.0°C to 175.0°C?**
    The specific heat of Al is 0.900 J/g °C.
    The correct answer is: 1.01 x 10⁴ J or 10.1 kJ.

75.0g q=SH·m·ΔT 175

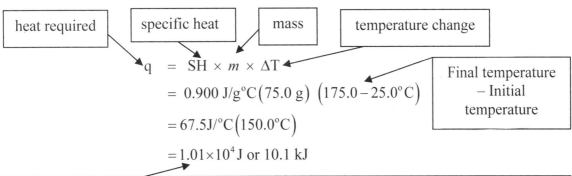

heat required   specific heat   mass   temperature change

$$q = SH \times m \times \Delta T$$
$$= 0.900 \, J/g°C(75.0 \, g)\,(175.0 - 25.0°C)$$
$$= 67.5 J/°C(150.0°C)$$
$$= 1.01 \times 10^4 \, J \text{ or } 10.1 \, kJ$$

Final temperature – Initial temperature

The heat required is positive indicating that the Al **absorbs** the heat.

CAUTION   When calculating ΔT use $T_{final} - T_{initial}$. This will insure that the sign of q is correct.

9

Module 2
Atomic Theory, Periodicity,
Chemical Bonding and Inorganic Nomenclature

Introduction

This module describes some basic concepts of atomic theory, periodicity, chemical bonding and inorganic nomenclature. The goals of this module are to show you how to:

1. discern the correct atomic electronic structure of the first 18 elements by looking at the periodic table
2. estimate atomic radii
3. estimate ionization energies
4. estimate electronegativities
6. determine if a compound is ionic or covalent
7. draw Lewis dot structures of atoms
8. write formulas of the simple ionic compounds
9. draw Lewis dot structures of ionic and covalent compounds
10. recognize if a covalent bond is polar or nonpolar
11. recognize resonance structures
12. use the rules of chemical nomenclature

You will need access to a periodic table in order to do the sample exercises in this module.

Module 2 Key Concepts

1. **Quantum numbers necessary to describe electrons in the elements H to Ar**

   $n = 1, 2, 3,$ and 4

   $\ell = s, p_x, p_y,$ or p_z

   $m_s = +1/2$ or $-1/2$

2. **Atomic radii**

   Atomic radii are the measured distances from the center of the atom to its outermost electrons.

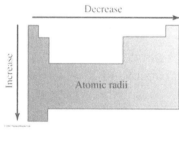

---

3. Ionization Energy

Ionization energy is the amount of energy required to remove an electron from an atom or ion. Ionization energy is an important indicator of an element's likelihood of forming positive ions.

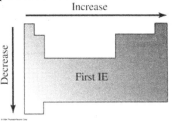

4. Electronegativity

Electronegativity is the relative measure of an element's ability to attract electrons to itself in a chemical compound. This property helps us determine the likelihood of ionic or covalent bond formation and the polarity of molecules.

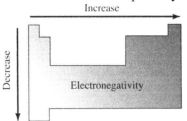

5. Nomenclature of Simple Ionic and Covalent Compounds

Ionic compounds are metal positive ions combined with:

a. **Nonmetal negative ions** (simple binary ionic compounds)
Nomenclature is the metal's name followed by nonmetal's stem plus –ide. If the metal positive ion is a transition metal, then add the oxidation state in parentheses after the metal's name.

b. **Polyatomic negative ions** (pseudobinary ionic compounds)
Nomenclature is metal's name followed by polyatomic ion's name.
Consult your textbook for a list of common polyatomic ions whose names and formulas you should recognize.

Binary Covalent compounds are composed of two nonmetals.
The less electronegative element is named first, and the more electronegative element is named second using stem plus –ide. Prefixes such as di-, tri-, etc. are used for both elements.

Sample Exercises

Proton, Neutron, and Electron Numbers in a Positive Ion

1. How many protons, neutrons, and electrons are present in one ion of this species?
$$^{48}Ca^{2+}$$

The correct answer is: $^{48}Ca^{2+}$ contains 20 protons, 28 neutrons, and 18 electrons.

Ca is element number 20 on the periodic table. This indicates that a Ca atom has 20 protons. In the symbol $^{48}Ca^{2+}$, the 48 is the sum of the protons and neutrons in this atom. Thus the number of neutrons is 48 – 20 protons = 28 neutrons. The symbol $^{48}Ca^{2+}$ indicates that this is an ion with a 2+ charge. Positively charged ions have lost electrons. Thus the $^{48}Ca^{2+}$ ion has lost two of its original 20 electrons leaving 18 electrons.

> **CAUTION**
>
> 1. On the periodic table, the uppermost number equals the number of protons in that element.
> 2. In an isotopic symbol, ^{48}Ca for example, the left superscript number is the sum of the protons and neutrons in the isotope.
> 3. For neutral atoms, the number of electrons is equal to the number of protons.
> 4. In an ionic symbol, Ca^{2+} for example, the right superscript number is the ionic charge. Positive ions have lost the number of electrons equal to the positive charge. Negative ions have gained the number of electrons equal to the negative charge.

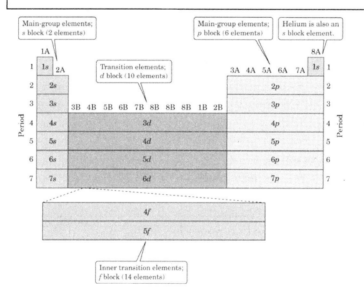

Principal Quantum Number, n

2. What is the value of the principal quantum number, n, for the outermost electrons in a Mg atom?

The correct answer is: n = 3

| Mg is on the 3rd row of the periodic chart. All of the elements on the 3rd row have n = 3. The value of n equals the row number except for the d-transition metals, lanthanides, and actinides. |

Each box on the periodic table represents one electron. You can use this along with the knowledge that each orbital holds two electrons to help you remember which part of the periodic table represents each type of orbital.

- The two electrons represented by the alkali and alkaline earth metal on each row of the periodic table represent one s orbital in each period.
- The six electrons represented by the elements boron through neon in row two represent three p orbitals (p_x, p_y, and p_z). The same is true for each row below period two.

Orbital Quantum Number, ℓ

3. *What is the ℓ value for the outermost electrons in a Mg atom?*

The correct answer is: ℓ = s electrons

Atoms in the IA and IIA columns of the periodic table have their outermost electrons in s orbitals. For the Mg atom these are 3s electrons indicating both their n and ℓ values.

4. *What is the ℓ value for the outermost electrons in a P atom?*

The correct answer is: ℓ = s and p electrons

Atoms in the IIIA to VIIIA columns of the periodic table are filling p electron orbitals. For the P atom these are 3p electrons indicating both their n and ℓ values. A P atom has two 3s electrons and three 3p electrons.

| You must think of quantum numbers as labels rather than numbers. |

Spin Quantum Number, m_s

5. *What is the m_s value for the outermost electrons in a Sr atom?*

The correct answer is: m_s = +1/2 and -1/2

| m_s can only have two possible values, +1/2 and -1/2. |

Electronic Structure from the Periodic Chart

6. *What is the correct electron configuration of the Mg atom? Write the configuration in both simplified notation and orbital box diagrams.*

The correct answer is: $1s^2\ 2s^2\ 2p^6\ 3s^2$ or [Ne] $3s^2$ - simplified notation

$$\boxed{\uparrow\downarrow}\quad\boxed{\uparrow\downarrow}\ \boxed{\uparrow\downarrow}\ \boxed{\uparrow\downarrow}\ \boxed{\uparrow\downarrow}\quad\boxed{\uparrow\downarrow}$$ **- orbital box diagram**

$1s$ $2s$ $2p_x$ $2p_y$ $2p_z$ $3s$

The noble gas core configuration is determined by starting at the element and decreasing the atomic number until reaching a noble gas.

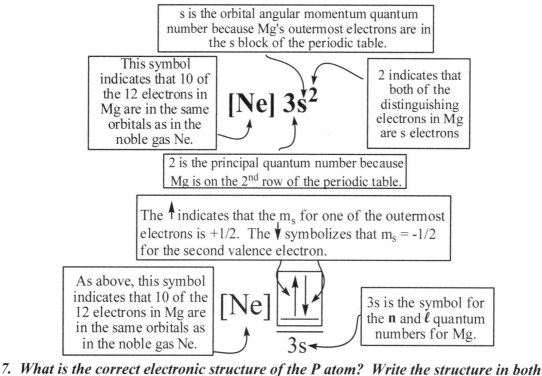

s is the orbital angular momentum quantum number because Mg's outermost electrons are in the s block of the periodic table.

This symbol indicates that 10 of the 12 electrons in Mg are in the same orbitals as in the noble gas Ne.

[Ne] 3s²

2 indicates that both of the distinguishing electrons in Mg are s electrons

2 is the principal quantum number because Mg is on the 2ⁿᵈ row of the periodic table.

The ↑ indicates that the m_s for one of the outermost electrons is +1/2. The ↓ symbolizes that m_s = -1/2 for the second valence electron.

As above, this symbol indicates that 10 of the 12 electrons in Mg are in the same orbitals as in the noble gas Ne.

[Ne] ↑↓
3s

3s is the symbol for the **n** and ℓ quantum numbers for Mg.

7. *What is the correct electronic structure of the P atom? Write the structure in both simplified notation and orbital box diagrams.*
   The correct answer is: [Ne] 3s² 3p³ or [Ne] ↑↓ ↑ ↑ ↑
                                                    3s 3pₓ 3p_y 3p_z

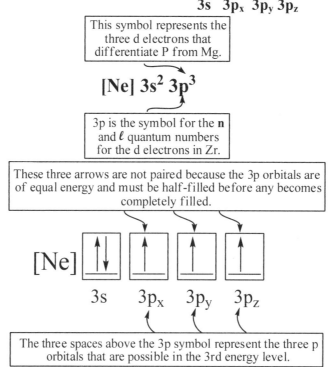

This symbol represents the three d electrons that differentiate P from Mg.

[Ne] 3s² 3p³

3p is the symbol for the **n** and ℓ quantum numbers for the d electrons in Zr.

These three arrows are not paired because the 3p orbitals are of equal energy and must be half-filled before any becomes completely filled.

[Ne] ↑↓ ↑ ↑ ↑
         3s 3pₓ 3p_y 3p_z

The three spaces above the 3p symbol represent the three p orbitals that are possible in the 3rd energy level.

TIPS — If the m_s values were reversed, the answers for both Mg and P would still be correct.

CAUTION — Three important rules that you need to know to understand electron configurations are 1) **orbitals fill in the order of increasing energy**, 2) **each orbital can hold up to two electrons with spins paired**, and 3) **when there is a set of orbitals of equal energy, each orbital becomes half-filled before any of them becomes completely filled**. Be certain that you know and understand these rules. One important thing to learn is how to determine the correct electron configuration of an element from the periodic table.

Atomic Radii

8. Arrange these elements by increasing atomic radii: F, Na, B, N

   The correct answer is: F < N < B < Na

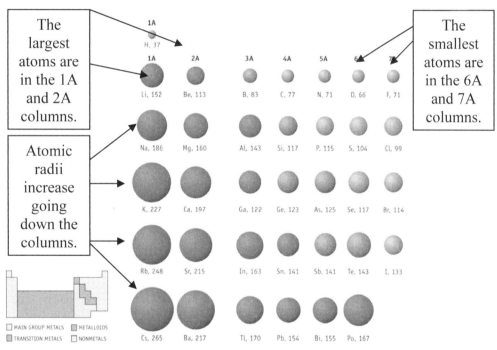

The largest atoms are in the 1A and 2A columns.

Atomic radii increase going down the columns.

The smallest atoms are in the 6A and 7A columns.

Ionization Energy

9. Arrange these elements by increasing ionization energies: F, N, C, O

   The correct answer is: C < O < N < F

First ionization energies increase steadily from the alkali metals to the noble gases.

15

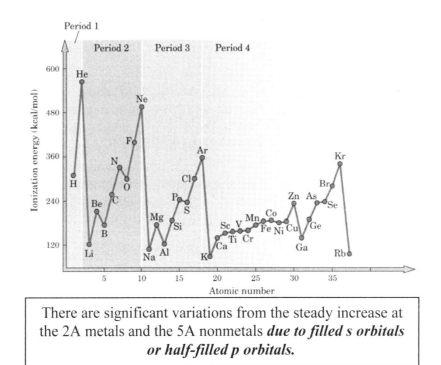

There are significant variations from the steady increase at the 2A metals and the 5A nonmetals *due to filled s orbitals or half-filled p orbitals.*

Electronegativity

10. Arrange these elements in the order of increasing electronegativity: O, Ca, Si, F
   The correct answer is: Ca < Si < O < F

The most electronegative elements are in the upper right corner of the periodic chart.

Table 4.5   Electronegativity Values of the Elements (Pauling Scale)

1A																	
H 2.1	2A												3A	4A	5A	6A	7A
Li 1.0	Be 1.5												B 2.0	C 2.5	N 3.0	O 3.5	F 4.0
Na 0.9	Mg 1.2	3B	4B	5B	6B	7B		8B		1B	2B		Al 1.5	Si 1.8	P 2.1	S 2.5	Cl 3.0
K 0.8	Ca 1.0	Sc 1.3	Ti 1.5	V 1.6	Cr 1.6	Mn 1.5	Fe 1.8	Co 1.8	Ni 1.8	Cu 1.9	Zn 1.6	Ga 1.6	Ge 1.8	As 2.0	Se 2.4	Br 2.8
Rb 0.8	Sr 1.0	Y 1.2	Zr 1.4	Nb 1.6	Mo 1.8	Tc 1.9	Ru 2.2	Rh 2.2	Pd 2.2	Ag 1.9	Cd 1.7	In 1.7	Sn 1.8	Sb 1.9	Te 2.1	I 2.5
Cs 0.7	Ba 0.9	La 1.1	Hf 1.3	Ta 1.5	W 1.7	Re 1.9	Os 2.2	Ir 2.2	Pt 2.2	Au 2.4	Hg 1.9	Tl 1.8	Pb 1.8	Bi 1.9	Po 2.0	At 2.2

The least electronegative elements are in the lower left corner of the periodic chart.

Electronegativity steadily increases moving from the lower left to the upper right corners of the periodic chart.

16

Determine if a Compound is Ionic or Covalent

11. Indicate which of the following compounds are ionic in nature and which are covalent in nature.

$$CH_4, KBr, Ca_3N_2, Cl_2O_7, H_2SO_4, InCl_3$$

The correct answer is: ionic = KBr, Ca_3N_2 and $InCl_3$
covalent = CH_4, Cl_2O_7, and H_2SO_4

	Look for metallic elements! Ionic compounds are formed by the reaction of metallic elements with nonmetallic elements or the reaction of the ammonium ion, NH_4^+, with nonmetals. Covalent compounds are formed by the reaction of two or more nonmetals
TIP	

K is a metallic element.	Ca is a metallic element.	In is a metallic element.

KBr   **Ca_3N_2**   **$InCl_3$**

Br is a nonmetal.	N is a nonmetal	Cl is a nonmetal
C is a nonmetal.	Cl is a nonmetal.	H and S are nonmetals.

CH_4   **Cl_2O_7**   **H_2SO_4**

H is a nonmetal.	O is a nonmetal.	O is a nonmetal.

Lewis Dot Structures of Atoms

12. Draw the correct Lewis dot structure of these elements: Mg, P, S, Ar
   The correct Lewis dot structures are shown below.

	Use the periodic table to determine the number of outermost electrons from each element's group number! Your first step in drawing Lewis dot structures should *always* be to count the species' electrons.
TIPS	

The number next to each dot represents the order in which it was added to the structure. Essentially, each of the four sides of the element's symbol represents an orbital. One side represents an s orbital, and the remaining three represent p orbitals. The s orbital must be filled first, followed by the three p orbitals. Note that it does not matter where you start nor whether you proceed clockwise or counterclockwise in your counting procedure.

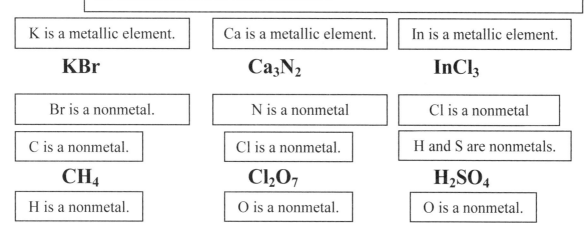

> Lewis dot structures reflect the electronic structures of the elements, including how the electrons are paired. Notice how the orbital diagrams match the Lewis dot structures of each element.

Table 2.8 Lewis Dot Structures for Elements 1–18 of the Periodic Table

1A	2A	3A	4A	5A	6A	7A	8A
H·							He:
Li·	Be:	B:	·C:	·N:	:O:	:F:	:Ne:
Na·	Mg:	Al:	·Si:	·P:	:S:	:Cl:	:Ar:

Each dot represents one valence electron.

Simple Ionic Compounds

13. Write the correct formulas of the ionic compounds formed when Mg atoms react with the following: a) Cl atoms, b) S atoms, c) P atoms.
The correct answers are: $MgCl_2$, MgS, and Mg_3P_2

> Like all of the 2A metals, Mg has two electrons in its outermost shell and commonly forms 2+ ions, Mg^{2+}.

$MgCl_2$	MgS	Mg_3P_2
Cl, and all of the VIIA nonmetals, have seven electrons in their outermost shell and commonly form 1- ions, Cl^-. Two Cl^- ions are required to balance the 2+ charge of the Mg and form a neutral compound.	S, and all of the VIA nonmetals, have six electrons in their outermost shell and commonly form 2- ions, S^{2-}. Only one S^{2-} ion is required to cancel out the 2+ charge on the Mg^{2+} ion.	P, and all of the VA nonmetals, have five electrons in their outermost shell and commonly form 3- ions, P^{3-}. Two P^{3-} ions are needed to balance the charge on three Mg^{2+} ions to form a neutral compound.

Drawing Lewis Dot Structures of Ionic Compounds

14. Draw the Lewis dot structures for each of these compounds: AlP, NaCl, $MgCl_2$
The correct structures are shown below.

When counting the outermost electrons, remember that a positive ion has, per positive charge, one electron less than the neutral parent atom. Per each negative charge, negative ions have one electron more than the neutral parent atom.

$$Al^{3+} \left[: \overset{\cdot\cdot}{P} : \right]^{3-} \qquad Na^+ \left[: \overset{\cdot\cdot}{Cl} : \right]^- \qquad Mg^{2+} \, 2\left[: \overset{\cdot\cdot}{Cl} : \right]^-$$

18

Al loses all three of its outermost electrons and forms a 3+ ion. Thus, it has no dots.

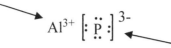

Al^{3+} $\begin{bmatrix} : \overset{\cdot\cdot}{\underset{\cdot\cdot}{P}} : \end{bmatrix}^{3-}$

P gains three electrons from Al, so it has 8 dots (5 outermost electron plus 3 from Al) and forms a 3- ion. The []'s indicate that the 3- charge is associated with the P ion.

Na loses its one outermost electron to form a 1+ ion, so it has no dots.

Na^{+} $\begin{bmatrix} : \overset{\cdot\cdot}{\underset{\cdot\cdot}{Cl}} : \end{bmatrix}^{-}$

Cl gains one electron from Na, thus has 8 dots (7 outermost electrons plus 1 from Na), and forms a 1- ion.

Mg loses both outermost electrons in forming a 2+ ion, so it has no dots.

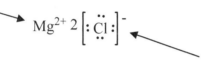

Mg^{2+} $2 \begin{bmatrix} : \overset{\cdot\cdot}{\underset{\cdot\cdot}{Cl}} : \end{bmatrix}^{-}$

Each Cl atom gains one electron from the Mg. The 2 in front of the []'s indicates that two Cl⁻ ions are needed to balance the charge of the Mg^{2+} ion.

Simple Covalent Compounds

15. Draw the correct Lewis dot structures for each of these compounds: SiH₄, PCl₃, SF₆.

The correct structures are:

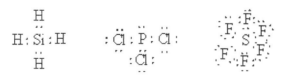

Try following these steps when drawing Lewis structures:

1. Determine the number of valence electrons in the compound.
2. Decide which atom is the central atom and make one bond (two electrons) to each of the remaining elements.
3. Complete the octet for all elements, and count how many electrons were used. Procedures to apply for exceptions to the octet rule will be discussed later in this module.

SiH_4 has 8 outermost electrons (4 from Si and 1 from each of the 4 H)

In most cases every element in a compound will obey the octet rule. Thus, Si has a share of 8 electrons and each H has a share of 2 electrons. This compound has only *bonding pairs* of electrons.

PCl_3 has 26 outermost electrons (5 from P and 7 from each of the 3 Cl).

In this compound P has a share of 8 electrons and each Cl has a share of 8 electrons. This compound has 3 *bonding pairs* and 1 *lone pair* of electrons.

SF_6 has 48 outermost electrons (6 from S and 7 from each of the 6 F).

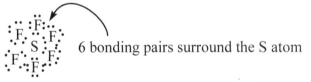

This compound does not obey the octet rule.
S has a share of 12 electrons while each F has a share of 8 electrons. This compound has 6 *bonding pairs* of electrons. Look in your textbook for the rules on which compounds do not obey the octet rule.

When drawing Lewis dot structures, if the compound obeys the octet rule, the central atom will have a share of 8 electrons. The possible combinations of 8 electrons for compounds that **obey the octet rule** are:

Bonding Pairs	Lone Pairs
4	0
3	1
2	2
1	3

If the compound **does not obey the octet rule**, the central atom can have 2, 3, 5 or 6 pairs of electrons around the central atom in some combination of bonding and lone pairs.

Noncentral atoms will obey the octet rule having either 1 bonding pair, as for H atoms, or a share of 8 electrons as is the case for Cl and F in the examples above.

Compounds Containing Multiple Bonds

16. Draw the correct Lewis dot structures for each of these compounds: CO_2 and N_2
   The correct structures are:

$$:\overset{..}{O} :: C :: \overset{..}{O}:$$

$$:N :: N:$$

CO_2 has 16 outermost electrons (4 from C and 6 from each of the 2 O).
C will be the central atom. Connecting each O with the central C by one bonding pair and filling in all octets results in the following structure:

$$:\overset{..}{\underset{..}{O}} \cdot\cdot \overset{..}{\underset{..}{C}} \cdot\cdot \overset{..}{\underset{..}{O}}:$$

Note that this structure contains 20 electrons, which is four more than the structure should have.

In order to decrease the number of electrons in a Lewis structure, make a double bond. *When the double bond is made one lone pair of electrons must be removed from each atom involved in the double bond.*

Create a double bond here, and remove the lone pairs that are circled.

$$:\overset{..}{O} :: \overset{..}{\underset{..}{C}} \cdot\cdot \overset{..}{\underset{..}{O}}:$$

This structure still has too many electrons, so the process is repeated on the other side of the molecule

N_2 has 10 outermost electrons. Connecting the two atoms and filling each octet results in a structure with 14 electrons:

$$:\overset{..}{\underset{..}{N}} \cdot\cdot \overset{..}{\underset{..}{N}}:$$

The formation of each multiple bond reduces the total electron count by 2 electrons. In this case, the process must be carried out twice in order to remove four electrons. The result is a *triple bond*.

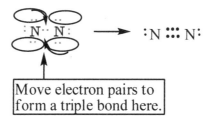

Move electron pairs to
form a triple bond here.

Resonance Structures

17. Draw the Lewis structure of the sulfur trioxide molecule, SO₃.

The correct answer is:

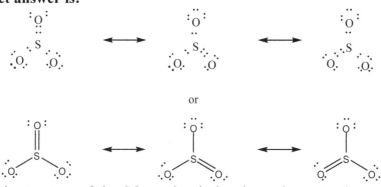

Once the Lewis structure of the SO_3 molecule has been drawn to the point that the O atoms have octets and the bonds to the S atom have been constructed, the structure looks like this:

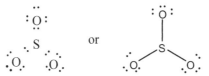

The S atom does not have an octet but all of the electrons have been used. To complete S's octet a double bond must be formed by moving one of the electron pairs from an O atom to the S-O bond. However, which O atom is the correct one? Since they are equivalent, the answer is none of them is preferred. This is an example of a molecule that exhibits resonance, a hybrid of two or more Lewis structures. To illustrate resonance one of the resonance hybrids formed by the movement of an electron pair is shown below:

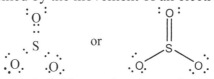

The three equivalent resonance hybrids can be shown by moving electrons (illustrated with curved arrows) in the following fashion.

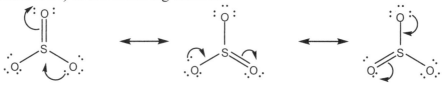

22

> **When writing resonance structures obey these rules:**
> 1. Look for a molecule requiring a multiple bond but more than one position on the molecule for the multiple bond is possible.
> 2. The atoms in the molecule cannot be moved but the electron pairs can.
> 3. Obey the rules for drawing Lewis dot structures.
> 4. The number of outermost electrons and the number of paired and unpaired electrons in each structure must be equal.

Polar or Nonpolar Covalent Bonds in Compounds
18. Which of these compounds contains polar covalent bonds?
$$F_2, CH_4, H_2O$$
The correct answer is: CH_4 and H_2O contain polar covalent bonds and F_2 does not.

The periodic trends regarding electronegativity are discussed above. You must understand those trends for problems of this nature.

Polar covalent bonds occur when the two atoms involved in the bond have a difference in electronegativity. In F_2 the two atoms are both F. They have the same electronegativity; thus, there is not a polar bond. In CH_4 and H_2O, the H to the central atom (C or O) bond involves atoms with different electronegativities. Thus there are polar covalent bonds in CH_4 and H_2O.

> Polar bonds have **dipoles** resulting from the *partial positive* and *partial negative* charges on atoms resulting from the unequal sharing of electrons. Dipoles are indicated by drawing an arrow over the bond with the head of the arrow pointing in the direction of the more electronegative atom. The H-Cl bond in HCl is polar, as shown below:
>
>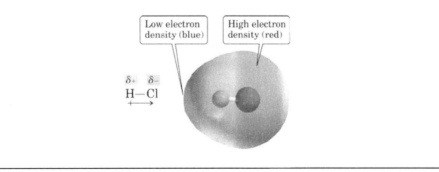

Simple Chemical Nomenclature
19. What is the correct name of the chemical compound $CaBr_2$?
The correct answer is: calcium bromide

Metal positive ions and nonmetal negative ions make simple binary ionic compounds. Simple binary ionic compounds are named using the metal's name followed by the nonmetal's stem and the suffix –ide. Prefixes like di- or tri- are **not used** to denote the number of ions present in the substance.

The metal positive ion in this case is Ca^{2+}, the calcium ion. The negative ion is Br^-, from the element bromine, whose ending is changed to –ide.

20. What is the correct name of the chemical compound $Mg_3(PO_4)_2$?
   The correct answer is: magnesium phosphate

Metal positive ions and polyatomic negative ions make <u>pseudobinary ionic compounds</u>. These compounds are named using the metal's name followed by the correct name of the polyatomic negative ion. Your textbook has a list of the polyatomic negative ions that you are expected to know. Make sure that you have the name, the negative ion's formula, and the charge memorized. Once again, no prefixes are used in these compounds to tell the number of ions present. Mg^{2+} is a positive ion made from the metal magnesium. PO_4^{3-} is a negative polyatomic ion named phosphate.

21. What is the correct name of this chemical compound, $FeCl_3$?
   The correct answer is: iron (III) chloride

Transition metal positive ions and nonmetal or polyatomic negative ions make <u>transition metal ionic compounds</u>. Their names are derived from the metal's name followed by the metal's oxidation state in Roman numerals inside parentheses. A metal's oxidation state is determined from the oxidation state of the negative ion. Fe^{3+} is a positive ion made from a transition metal (B Groups on the periodic chart). Cl^{1-} is a negative ion made from the nonmetal chlorine.

22. What is the correct name of this chemical compound, N_2O_4?
   The correct answer is: dinitrogen tetroxide

This compound is made from two nonmetals, nitrogen and oxygen, so it is a <u>binary covalent compound</u>. *These compounds use prefixes to indicate the number of atoms of each element present in the compound.* This is an important difference from the ionic compounds in the previous examples.

△ TIPS	1. If the compound contains a metal cation, then you must **<u>NOT</u>** use prefixes in the name. Prefixes are used only in covalent compounds.
	2. Check the location of the metal on the periodic table. If the metal is a transition metal, then you likely need to use a Roman numeral to indicate its oxidation state.
	3. Be sure that you are familiar with the names and formulas of the common polyatomic ions.

Module 3
Nuclear Chemistry

Introduction

This module discusses the basic relationships used in a typical nuclear chemistry chapter. The important topics described include:

1. predicting the products of alpha, negatron, and positron radioactive decays as well as of nuclear reactions
2. nuclear half-lives
3. radiation dosimetry
4. fission reactions
5. fusion reactions

Module 3 Key Concepts

1. $^A_Z X \rightarrow\ ^{A-4}_{Z-2} Y + ^4_2 He$

   This is the basic equation for radioactive alpha decay. Alpha decay removes two protons and two neutrons, in the form of a ^4He nucleus, from the decaying nucleus converting the element X into a new element Y.

2. $^A_Z X \rightarrow\ ^A_{Z+1} Y + ^0_{-1} e$ (or $^0_{-1}\beta^-$)

   Radioactive beta decay, β^- or negatron decay, converts a neutron into a proton by eliminating a high velocity electron, the β^- particle, from the nucleus.

   $^A_Z X \rightarrow\ ^A_{Z-1} Y + ^0_{+1} e$ (or $^0_{+1}\beta^+$)

   Radioactive positron decay, β^+, converts a proton into a neutron by eliminating a high velocity positive electron, the β^+ particle, from the nucleus.

3. $^A_Z X + ^0_{-1} e \rightarrow\ ^A_{Z-1} Y$

   Radioactive electron capture decay converts a proton into a neutron by catching an electron inside the nucleus.

4. $^A_Z X \rightarrow\ ^A_Z X + \gamma$

   Radioactive gamma decay converts a nucleus in an excited state into a nucleus in a lower energy state by emitting a high energy light particle called a gamma (γ) ray.

5. $^{M_1}_{Z_1} Q \rightarrow\ ^{M_2}_{Z_2} R + ^{M_3}_{Z_3} Y$ where $M_1 = M_2 + M_3$ and $Z_1 = Z_2 + Z_3$

   This is the basic relationship for nuclear reactions, and radioactive decays. The proton numbers of the product nuclides (Z_2 and Z_3) must sum to the original nuclide's proton number, Z_1. The mass numbers of the product nuclides (M_2 and M_3) must also add up to the original nuclide's mass, M_1.

6. **half-life, $t_{1/2}$, time necessary for ½ of the starting material to radioactively decay**

   A 100.0 g sample of a radioactive substance with a $t_{1/2} = 0.50$ hours will have 50.0 g of the sample remaining after 30 minutes.

Sample Exercises

Alpha Decay
1. What is the product nucleus of the alpha decay of ^{232}Th?
   The correct answer is: ^{228}Ra.

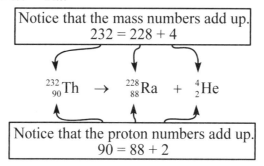

Notice that the mass numbers add up.
232 = 228 + 4

$$^{232}_{90}Th \rightarrow \,^{228}_{88}Ra \,+\, ^{4}_{2}He$$

Notice that the proton numbers add up.
90 = 88 + 2

TIPS

Alpha decay occurs primarily in nuclides that have more than 83 protons. To determine the product nuclide, take the proton number of the decaying nucleus and subtract 2. The product's mass number will be the decaying nuclide's mass number minus 4.

Beta Decay
2. What is the product nucleus of the β^-, negatron, decay of ^{14}C?
   The correct answer is: ^{14}N.

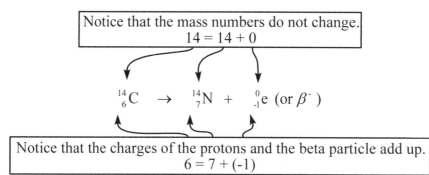

Notice that the mass numbers do not change.
14 = 14 + 0

$$^{14}_{6}C \rightarrow \,^{14}_{7}N \,+\, ^{0}_{-1}e \,(or\, \beta^-)$$

Notice that the charges of the protons and the beta particle add up.
6 = 7 + (-1)

3. What is the product nucleus of the β^+, positron, decay of ^{37}Ca?
   The correct answer is: ^{37}K.

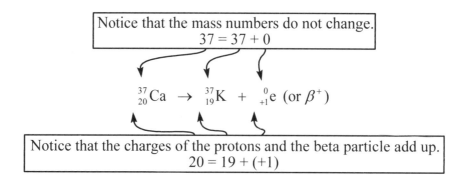

Notice that the mass numbers do not change.
37 = 37 + 0

$$^{37}_{20}Ca \rightarrow \,^{37}_{19}K \,+\, ^{0}_{+1}e \,(or\, \beta^+)$$

Notice that the charges of the protons and the beta particle add up.
20 = 19 + (+1)

4. What is the product nucleus of the electron capture decay of ^{37}Ar?
   The correct answer is: ^{37}Cl.

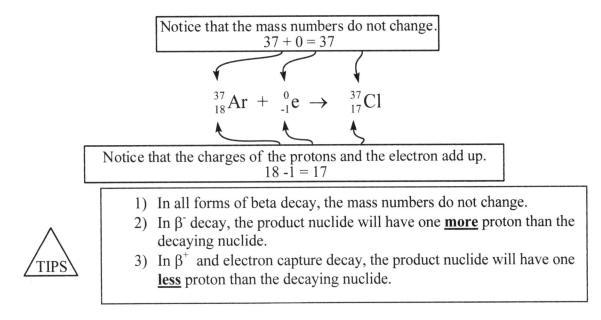

Notice that the mass numbers do not change.
$37 + 0 = 37$

$$^{37}_{18}Ar + ^{0}_{-1}e \rightarrow ^{37}_{17}Cl$$

Notice that the charges of the protons and the electron add up.
$18 - 1 = 17$

TIPS

1) In all forms of beta decay, the mass numbers do not change.
2) In β^- decay, the product nuclide will have one **more** proton than the decaying nuclide.
3) In β^+ and electron capture decay, the product nuclide will have one **less** proton than the decaying nuclide.

Gamma Decay
5. What is the product nucleus of the gamma decay of ^{23}Na?
   The correct answer is: ^{23}Na.

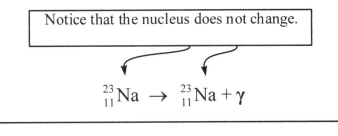

Notice that the nucleus does not change.

$$^{23}_{11}Na \rightarrow ^{23}_{11}Na + \gamma$$

In gamma decay, the nucleus transitions from an excited nuclear state to a lower energy by emitting a high energy photon.

TIPS

In all gamma decay, the starting nucleus and the product nucleus are the same. There is no change in the proton or neutron numbers.

Nuclear Reactions
6. Fill in the missing nuclide in this nuclear reaction.

$$^{53}Cr + ^{4}He \rightarrow \underline{\hspace{1cm}} + 2n$$

The correct answer is: ^{55}Fe.

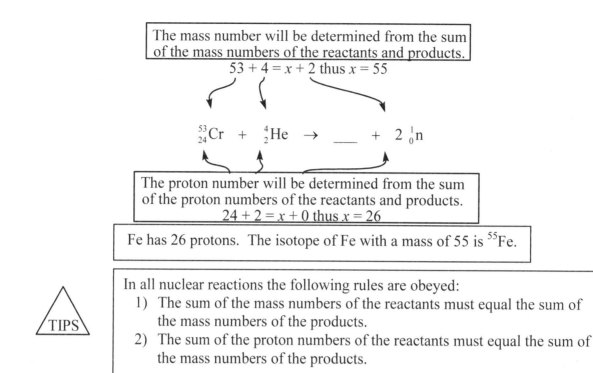

The mass number will be determined from the sum of the mass numbers of the reactants and products.
$53 + 4 = x + 2$ thus $x = 55$

$$^{53}_{24}\text{Cr} \ + \ ^{4}_{2}\text{He} \ \rightarrow \ \_\_\_ \ + \ 2\,^{1}_{0}\text{n}$$

The proton number will be determined from the sum of the proton numbers of the reactants and products.
$24 + 2 = x + 0$ thus $x = 26$

Fe has 26 protons. The isotope of Fe with a mass of 55 is ^{55}Fe.

TIPS

In all nuclear reactions the following rules are obeyed:
1) The sum of the mass numbers of the reactants must equal the sum of the mass numbers of the products.
2) The sum of the proton numbers of the reactants must equal the sum of the mass numbers of the products.

Half-lives, $t_{1/2}$, for Radioactive Decay

7. Tritium, ^{3}H, a radioactive isotope of hydrogen has a half-life of 12.26 y. If 2.0 g of ^{3}H were made, how much of it would be left 36.78 y later?
The correct answer is: 0.25 g.

$$2.0 \text{ g of } ^{3}\text{H} \xrightarrow{12.26 \text{ y}} 1.0 \text{ g of } ^{3}\text{H in the 1}^{\text{st}} \ t_{1/2}$$

$$1.0 \text{ g of } ^{3}\text{H} \xrightarrow[24.52 \text{ y total}]{\text{another 12.26 y}} 0.50 \text{ g of } ^{3}\text{H in the 2}^{\text{nd}} \ t_{1/2}$$

$$0.50 \text{ g of } ^{3}\text{H} \xrightarrow[36.78 \text{ y total}]{\text{another 12.26 y}} 0.25 \text{ g of } ^{3}\text{H in the 3}^{\text{rd}} \ t_{1/2}$$

8. A sample of a radiopharmaceutical containing 10.0 mg of ^{15}O is injected into a patient. A few minutes later the patient's body contains 0.625 mg of ^{15}O still remaining from the sample. How long has the radiopharmaceutical been inside the patient? The half-life of ^{15}O is 2.03 minutes.
The correct answer is: 8.12 minutes

0.625 mg of radiopharmaceutical remaining in the patient from the original 10.0 mg indicates that one-sixteenth of the sample remains, $\left(\dfrac{0.625 \text{ mg}}{10.0 \text{ mg}} = 0.0625 = \dfrac{1}{16} \right)$. To decrease the original sample by one-sixteenth the sample must have passed through four half-lives, $\left(\dfrac{1}{2} \times \dfrac{1}{2} \times \dfrac{1}{2} \times \dfrac{1}{2} = \dfrac{1}{16} \right)$. Thus the radiopharmaceutical has been inside the patient for 4×2.03 minutes $= 8.12$ minutes.

Radiation Dosimetry

9. Which of the following units of radiation dosimetry provides the most useful information regarding a patient's radiation exposure from a radiopharmaceutical?
Roentgen, rad, rem

The correct answer is: rem

The Roentgen, a measure of the radiation emitted by a radioactive source, tells us the amount of radiation emitted by a source but does not indicate its effect on a human. The rad, a measure of the radiation absorbed by the tissue versus that delivered to the tissue, indicates how much radiation is absorbed by the body but does not differentiate between the damage done by alpha, beta, or gamma radiation. The Rem (an acronym for Roentgen equivalent man), a measure of the tissue damage done by radiation exposure, correctly indicates the damage inflicted by all types of radiation. A Sievert (Sv) is the metric system unit similar to the Rem. A Sievert = 100 rem. This table shows the effects of γ radiation exposure to humans.

Radiation Exposure	Effect
< 25 rem	None
25 to 100 rem	Short-term reduction in blood cells
100 – 200 rem	Nausea, fatigue, vomiting for > 125 rem, longer term blood cell reduction
200 - 300 rem	First day nausea and vomiting, two weeks later appetite loss, sore throat, diarrhea, death for 10-35% of humans within 30 days, remaining recover in ~ 3 months
300 - 600 rem	Within hours nausea, vomiting, and diarrhea, one week later bleeding and inflammation of mouth and throat, for > 450 rems 50% of people die
> 600 rem	Within hours nausea, vomiting, and diarrhea, followed by death in two weeks for nearly 100% of humans

Fission and Fusion Reactions

10. What is the product nucleus of this fission reaction?

$$^{239}Pu + n \rightarrow \, ^{100}Nb + \, ? \, + 2n$$

Correct answer is: ^{137}I

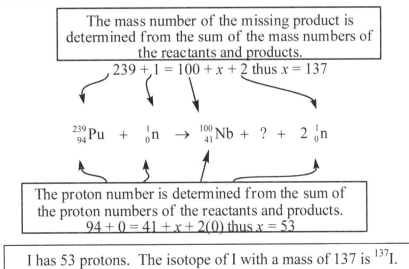

The mass number of the missing product is determined from the sum of the mass numbers of the reactants and products.

$239 + 1 = 100 + x + 2$ thus $x = 137$

$$^{239}_{94}Pu \, + \, ^{1}_{0}n \, \rightarrow \, ^{100}_{41}Nb + \, ? \, + \, 2\,^{1}_{0}n$$

The proton number is determined from the sum of the proton numbers of the reactants and products.

$94 + 0 = 41 + x + 2(0)$ thus $x = 53$

I has 53 protons. The isotope of I with a mass of 137 is ^{137}I.

11. What is the product nucleus of this fusion reaction?

$$^4He + {^3}H \rightarrow \ ? \ + 2n$$

Correct answer is: ^{6}Li

> Again, the mass number of the missing product is determined from the sum of the mass numbers of the reactants and products.

$$4 + 3 = x + 1 \text{ thus } x = 6$$

$$^4_2He \ + \ ^3_1H \ \rightarrow \ ? \ + \ ^1_0n$$

> The missing proton number is determined from the sum of the proton numbers of the reactants and products.

$$2 + 1 = x + 0 \text{ thus } x = 3$$

> Li has 3 protons. The isotope of Li with a mass of 6 is ^{6}Li.

> In **fission** reactions *a more massive nucleus is split into two lighter nuclei.* A representation of a chain fission reaction showing four completed fission reactions is shown below.
>
> In **fusion** reactions *two lighter nuclei are merged into a more massive nucleus.* Above, the reaction indicates that a ^{4}He and ^{3}H merge into a ^{6}Li nucleus. This is one possible fusion reaction that occurs in stars.

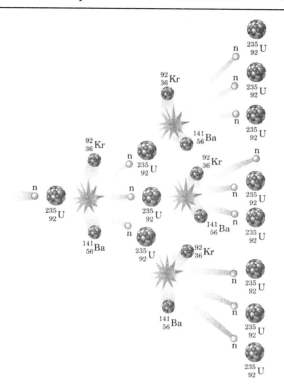

Module 4
Chemical Reactions and Stoichiometry

Introduction

In this module we will look at several problems that involve chemical reactions and reaction stoichiometry. The important points to learn in this module are:

1. how to interpret, use, and perform all of the important calculations that involve the mole
2. how to balance chemical reactions
3. basic reaction stoichiometry
4. limiting reactant calculations
5. percent yield calculations
6. how to use the reactants of a chemical reaction to determine the reaction type
7. predicting the reaction products of metathesis reactions

You will need a periodic table to calculate molecular weights in these problems.

Module 4 Key Concepts

1. **Molar mass** $= \sum$ **atomic weights of atoms in a compound, molecule, or ion**

   The molar mass, molecular weight, or formula weight[*] is calculated by summing the atomic weights of the atoms in the compound. This value gives the mass in grams of one mole of a substance.

2. **One mole = 6.022 x 10^{23} particles**

   Avogadro's relationship is used to convert from the number of moles of a substance to the number of atoms, ions, or molecules of that substance and vice versa.

3. **mass of one atom of an element** $= \left(\dfrac{\textbf{grams of an element}}{\textbf{1 mole of an element}} \right) \left(\dfrac{\textbf{1 mole of atoms}}{\textbf{6.022} \times \textbf{10}^{23} \textbf{ atoms}} \right)$

   The mass of one atom, ion, or molecule is used to determine the mass of a few atoms, ions, or molecules of a substance. Notice that the fraction in the first set of parentheses simply represents molar mass

4. **Mole ratio**

   The chemical formula of a compound indicates the *ratio* of the different types of atom in the compound. The mole ratio can be used to convert from mass or moles of a compound to mass or moles of a specific atom in the compound.

5. **Percent yield**

   $$\% \text{ yield} = \frac{\text{actual yield}}{\text{theoretical yield}} \times 100$$

   The percent yield formula is used to determine the percentage of the theoretical yield that was formed in a reaction.

---

*The terms molar mass, molecular weight, and formula weight all apply to the same concept/calculation. Technically, the term molecular weight should be used only with covalent compounds and formula weight applies only to ionic compounds. The more generic term *molar mass* is used frequently in chemical literature.

<u>Sample Exercises</u>
Determining the Molar Mass
1. What is the molar mass (formula weight) of calcium phosphate, Ca₃(PO₄)₂?
   The correct answer is: 310.2 g/mol

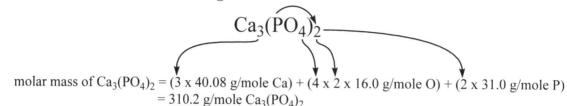

molar mass of $Ca_3(PO_4)_2$ = (3 x 40.08 g/mole Ca) + (4 x 2 x 16.0 g/mole O) + (2 x 31.0 g/mole P)
                            = 310.2 g/mole $Ca_3(PO_4)_2$

Determining the Number of Moles
2. How many moles of calcium phosphate are there in 65.3 g of Ca₃(PO₄)₂?
   The correct answer is: 0.211 mol Ca₃(PO₄)₂

$$? \text{ moles of } Ca_3(PO_4)_2 = 65.3 \text{ g } Ca_3(PO_4)_2 \left(\frac{1 \text{ mol } Ca_3(PO_4)_2}{310.2 \text{ g } Ca_3(PO_4)_2} \right)$$
$$= 0.211 \text{ mol } Ca_3(PO_4)_2$$

Molar mass of calcium phosphate from exercise #1.

Once the number of moles of the sample is known, we can determine the number of molecules or formula units of the substance. (Molecules are found in covalent compounds. Ionic compounds do not have molecules thus their smallest subunits are named formula units.)

Determining the Number of Molecules or Formula Units
3. How many formula units of calcium phosphate are there in 0.211 moles of Ca₃(PO₄)₂?

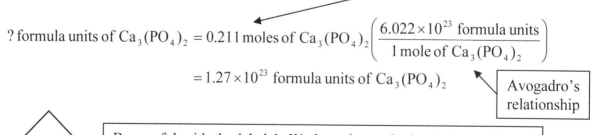

$$? \text{ formula units of } Ca_3(PO_4)_2 = 0.211 \text{ moles of } Ca_3(PO_4)_2 \left(\frac{6.022 \times 10^{23} \text{ formula units}}{1 \text{ mole of } Ca_3(PO_4)_2} \right)$$
$$= 1.27 \times 10^{23} \text{ formula units of } Ca_3(PO_4)_2$$

Avogadro's relationship

⟨CAUTION⟩ Be careful with the labels! We have just calculated the number of *formula units*. Do not confuse this with the number of atoms or the number of ions! All are valid questions with different answers!

Determining the Number of Atoms or Ions
4. How many oxygen, O, atoms are there in 0.211 moles of Ca₃(PO₄)₂?

The correct answer is: 1.02 x 10²⁴ oxygen atoms

Using the last idea from the key concepts box, we can determine the mass of a few molecules or formula units of a compound.

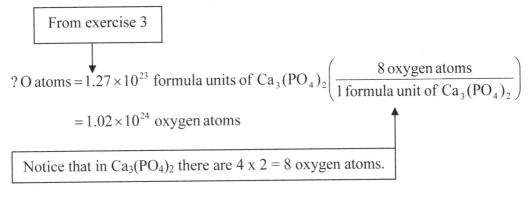

$$? \text{O atoms} = 1.27 \times 10^{23} \text{ formula units of } Ca_3(PO_4)_2 \left(\frac{8 \text{ oxygen atoms}}{1 \text{ formula unit of } Ca_3(PO_4)_2} \right)$$

$$= 1.02 \times 10^{24} \text{ oxygen atoms}$$

Notice that in $Ca_3(PO_4)_2$ there are 4 x 2 = 8 oxygen atoms.

Determining the Mass of Molecules or Formula Units of a Substance

5. *What is the mass, in grams, of 25.0 formula units of Ca₃(PO₄)₂?*
The correct answer is: 1.29 x 10⁻²⁰ g

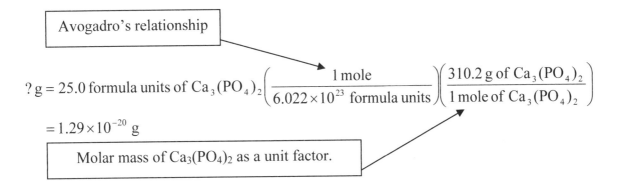

$$? \text{g} = 25.0 \text{ formula units of } Ca_3(PO_4)_2 \left(\frac{1 \text{ mole}}{6.022 \times 10^{23} \text{ formula units}} \right) \left(\frac{310.2 \text{ g of } Ca_3(PO_4)_2}{1 \text{ mole of } Ca_3(PO_4)_2} \right)$$

$$= 1.29 \times 10^{-20} \text{ g}$$

Molar mass of $Ca_3(PO_4)_2$ as a unit factor.

Combined Equations

6. *How many carbon, C, atoms are there in 0.375 g of C₄H₈O₂?*
The correct answer is: 1.03x 10²² atoms

molecular weight of $C_4H_8O_2$ Avogadro's relationship

$$? \text{C atoms} = 0.375 \text{g of } C_4H_8O_2 \left(\frac{1 \text{ mol of } C_4H_8O_2}{88.0 \text{ g of } C_4H_8O_2} \right) \left(\frac{6.022 \times 10^{23} \text{ molecules}}{1 \text{ mol of } C_4H_8O_2} \right) \left(\frac{4 \text{ carbon atoms}}{1 \text{ molecule of } C_4H_8O_2} \right)$$

$$= 1.03 \times 10^{22} \text{ C atoms}$$

The molecular formula indicates that there are 4 carbon atoms in every molecule of $C_4H_8O_2$.

> One of the problems most commonly encountered by students is determining where to start on these problems. If you have trouble getting started, focus on the information that was given. All of the examples up to this point in this module began by using the mass or number of moles stated in the question. You will almost always use some combination of molar masses, Avogadro's relationship, and mole ratio to solve these problems. Select which to use first by looking at the units of the information given and determining how to cancel them.

Pay attention to vocabulary! Keep in mind the differences between atoms, ions, and molecules, and pay attention to which pertains to the question asked.

Chemical Reactions

Chemical reactions symbolize what happens when chemical substances are mixed and new substances are formed. Before proceeding, it is important to review some vocabulary:

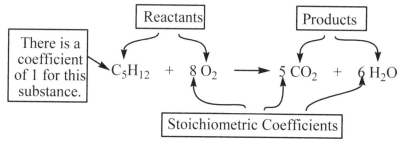

The chemical species which begin the reaction are _reactants_. _Products_ are the species resulting from the reaction. Stoichiometric coefficients are required to "_balance_" the equation. This process insures that equal numbers of atoms of each element are present on both sides of the reaction. Otherwise the reaction would violate the Law of Conservation of Mass.

> Equations can be balanced using several methods. While it does not matter which compound you balance first it is often easiest to use the following steps:
> 1. If possible, start with an element that appears in only one compound on each side of the equation.
> 2. Save balancing anything that appears without other elements (O_2, Fe(s), etc.) for last.
> 3. If the equation contains polyatomic ions, SO_4^{2-} for example, you may try balancing them as whole entities rather than balancing each individual element.

Sample Exercises
Balancing Chemical Reactions

7. *Balance this chemical reaction using the smallest whole numbers.*

$$Ca(OH)_2 + H_3PO_4 \rightarrow Ca_3(PO_4)_2 + H_2O$$

Consider starting with Ca since it appears in only one compound on each side of the reaction. It is probably easiest to balance (PO_4^{3-}) as the polyatomic ion rather than individually as P and O atoms. Then, only H and O are left.

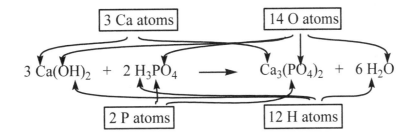

8. Balance this chemical reaction using the smallest whole numbers.

$$C_6H_{14} + O_2 \rightarrow CO_2 + H_2O$$

Start with C, and save oxygen to balance last!

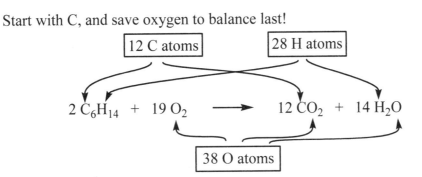

Simple Reaction Stoichiometry

Properly balanced chemical equations are _necessary_ for reaction stoichiometry calculations.

9. How many moles of H_2 can be formed from the reaction of 3.0 moles of Na with excess H_2O.
   The correct answer is: 1.5 moles H_2

$$2\,Na \;+\; 2\,H_2O \;\rightarrow\; 2\,NaOH \;+\; H_2$$

TIP

The word **excess** is important. It is your clue that this problem does **not** involve a limiting reactant calculation.

The reaction ratio, which comes from the balanced reaction, is 2 moles of Na consumed for every 1 mole of H_2 formed. Write it as a unit factor.

$$? \text{ moles } H_2 = 3.0 \text{ moles } Na \left(\frac{1 \text{ mole of } H_2}{2 \text{ moles of } Na} \right) = 1.5 \text{ moles } H_2$$

The reaction ratio is a conversion factor relating moles of any reactant or product to moles of another reactant or product. The reaction ratio comes from the balanced equation. (Some texts refer to the reaction ratio as the mole ratio.)

| TIP | The reaction ratio is the ONLY way to use information about one species in the reaction to determine something about a *different species* in the reaction. |

10. *How many grams of H_2 can be formed from the reaction of 11.2 grams of Na with excess H_2O?*
   The correct answer is: 0.494 g H_2

$$2\,Na \;+\; 2\,H_2O \;\rightarrow\; 2\,NaOH \;+\; H_2$$

| Converts g of Na to moles of Na | Reaction Ratio | Converts moles of H_2 to g of H_2 |

$$? \text{ grams of } H_2 = 11.2 \text{ g} \left(\frac{1 \text{ mole of } Na}{22.9 \text{ g of } Na} \right) \left(\frac{1 \text{ mole of } H_2}{2 \text{ moles of } Na} \right) \left(\frac{2.02 \text{ g of } H_2}{1 \text{ mole of } H_2} \right) = 0.494 \text{ g of } H_2$$

This problem makes the complete transformation from grams of one of the reactants, Na, to grams of one of the products, H_2. There is a very common set of transformations that are used in this calculation and which are used in many reaction stoichiometry problems.

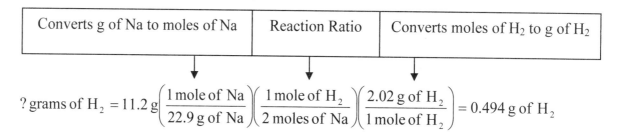

grams of X $\xrightarrow{\substack{\text{molar mass} \\ \text{of X}}}$ **moles of X** $\xrightarrow{\substack{\text{reaction} \\ \text{ratio}}}$ **moles of Y** $\xrightarrow{\substack{\text{molar mass} \\ \text{of Y}}}$ **grams of Y**

Notice how all of the units cancel, leaving g of H_2.

Limiting Reagents
11. *What is the maximum number of grams of H_2 that can be formed from the reaction of 11.2 grams of Na with 9.00 grams of H_2O?*
   The correct answer is: 0.494 g of H_2

$$2\,Na + 2\,H_2O \rightarrow 2\,NaOH + H_2$$

TIP

The word excess is not in this problem, and amounts of both reactants are given. These are your *clues that this is a limiting reagent problem.*

You must perform reaction stoichiometry steps for each reactant with an amount given in the problem. In this case that is two steps.

$$? \text{ grams of } H_2 = 11.2 \text{ g of Na} \left(\frac{1 \text{ mole of Na}}{22.9 \text{ g of Na}} \right) \left(\frac{1 \text{ mole of } H_2}{2 \text{ moles of Na}} \right) \left(\frac{2.02 \text{ g of } H_2}{1 \text{ mole of } H_2} \right) = 0.494 \text{ g of } H_2$$

$$? \text{ grams of } H_2 = 9.00 \text{ g of } H_2O \left(\frac{1 \text{ mole of } H_2O}{18.0 \text{ g of } H_2O} \right) \left(\frac{1 \text{ mole of } H_2}{2 \text{ moles of } H_2O} \right) \left(\frac{2.02 \text{ g of } H_2}{1 \text{ mole of } H_2} \right) = 0.505 \text{ g of } H_2$$

YIELD

The maximum amount will be the _smallest_ amount that you calculate in the reaction stoichiometry steps!

This calculation indicates that all 11.2 g of Na are used in the production of 0.494 g of H_2. Since there is no Na left, no more H_2 can be produced, even though there is still H_2O remaining. Once one reactant is completely used, no more product can be made. In this example, Na is the *limiting reactant* and H_2O is the *excess reactant*.

Percent Yield
12. If 11.2 g of Na reacts with 9.00 g of H_2O and 0.400 g of H_2 is formed, what is the percent yield of the reaction?
The correct answer is: 81.0%

This is the actual yield.

$$2 \, Na + 2 \, H_2O \rightarrow 2 \, NaOH + H_2$$

TIP

Key clues that indicate percent yield problems are: a) amounts of both reactants given, b) an amount for the product, and c) the words percent yield.

In percent yield problems, limiting reactant calculations are frequently performed first to determine the *theoretical yield*. This is the amount of product that is formed if the reaction goes 100% to completion (this rarely happens in the lab!) and what was calculated in exercises 4 and 5. For this problem the theoretical yield is the same as determined in exercise 5 (0.494 g of H_2).

$$\% \text{ yield} = \frac{\text{actual yield}}{\text{theoretical yield}} \times 100 = \frac{0.400 \text{ g}}{0.494 \text{ g}} \times 100\% = 81.0\%$$

Reduction-Oxidation Reactions

Reduction-oxidation reactions are those in which electrons are transferred from one species to another. Reductions cannot occur without accompanying oxidations, so these are often called *redox reactions*. Your textbook has a series of rules for assigning oxidation numbers to elements in chemical species. If you do not know the rules for oxidation states, learn them now.

13. What reaction type is represented by this chemical reaction?
$$2\,Na(s)\ +\ 2\,H_2O(\ell)\ \rightarrow\ 2\,NaOH(aq)\ +\ H_2(g)$$

The correct answer is: This is a redox reaction.

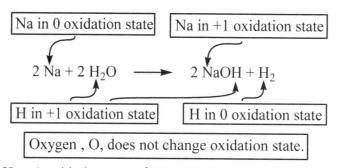

H: +1 oxidation state → 0 oxidation state = reduction
Na: 0 oxidation state → +1 oxidation state = oxidation

To recognize redox reactions you must *look for chemical species that are changing their oxidation states.*

Notice that Na in the above reaction is in its *elemental state* on the reactant side of the reaction and in a compound on the other side (the same is true of H). This is a big clue that you are dealing with a redox reaction. All species in their elemental states have oxidation states of zero, and species in compounds typically do not have oxidation states of zero. Thus, the oxidation state probably changes during the reaction!

Metathesis Reactions
14. What reaction type is represented by this reaction?
$$Ba(OH)_2(aq)\ +\ H_2SO_4(aq)\ \rightarrow\ BaSO_4(s)\ +\ 2\,H_2O(\ell)$$

The correct answer is: This is a metathesis reaction.

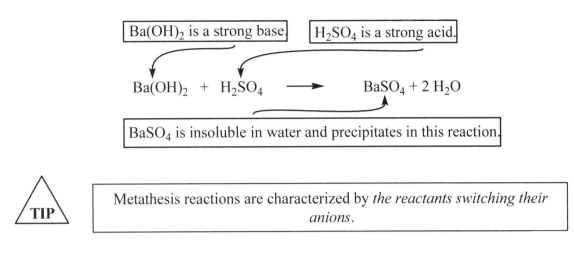

Ba(OH)$_2$ is a strong base. | H$_2$SO$_4$ is a strong acid.

$$Ba(OH)_2 \ + \ H_2SO_4 \ \longrightarrow \ BaSO_4 + 2\,H_2O$$

BaSO$_4$ is insoluble in water and precipitates in this reaction.

TIP | Metathesis reactions are characterized by *the reactants switching their anions.*

This is exhibited by using the symbols AB to represent one reactant and CD to represent the other reactant. The products are represented by AD and CB.

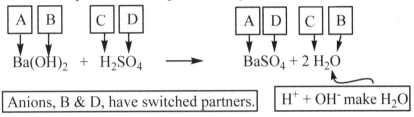

| A | B | | C | D | → | A | D | | C | B |
$$Ba(OH)_2 \ + \ H_2SO_4 \ \longrightarrow \ BaSO_4 + 2\,H_2O$$

Anions, B & D, have switched partners. | H$^+$ + OH$^-$ make H$_2$O

When an acid reacts with a base, both a salt (BaSO$_4$ in this case) and water (if the base is a hydroxide) will be formed. Water is formed by the combination of the H$^+$ with the OH$^-$ when the anions switch partners.

TIP | Precipitation reactions are characterized by *the formation of a compound that is insoluble in water.*

TIP | You must understand and use the solubility rules from your textbook to recognize a precipitation reaction since the phases of the product compounds are not frequently given, as you will see in the exercises below.

Predicting Products of Metathesis Reactions
15. What are the products of this chemical reaction?
$$Sr(OH)_2(aq) \ + \ Fe(NO_3)_3(aq) \rightarrow ??? + ???$$

The correct answer is: Sr(NO$_3$)$_2$ and Fe(OH)$_3$

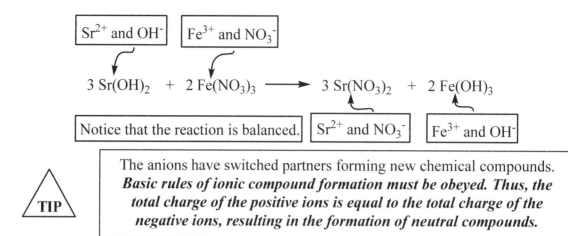

$$3 \text{ Sr(OH)}_2 \quad + \quad 2 \text{ Fe(NO}_3)_3 \quad \longrightarrow \quad 3 \text{ Sr(NO}_3)_2 \quad + \quad 2 \text{ Fe(OH)}_3$$

Sr^{2+} and OH$^-$

Fe^{3+} and NO$_3^-$

Notice that the reaction is balanced.

Sr^{2+} and NO$_3^-$

Fe^{3+} and OH$^-$

TIP

The anions have switched partners forming new chemical compounds. **_Basic rules of ionic compound formation must be obeyed. Thus, the total charge of the positive ions is equal to the total charge of the negative ions, resulting in the formation of neutral compounds._**

Module 5
States of Matter, Solutions and Colligative Properties

Introduction

This module describes the basic laws that govern the three states of matter: gas, liquid, and solid. The goals of the module are to:

1. become familiar with the combined, ideal and Dalton's gas laws
2. learn how to determine the relative freezing and boiling points of various liquids based on their intermolecular forces
3. learn to determine the relative melting points of various solids based on the strength of their bonding
4. become familiar with phase diagrams
5. predict species' solubility in various solvents
6. learn how to increase the solubility of a given species in a solvent
7. convert from one common concentration unit to another
8. perform dilution problems
9. determine the freezing point of solutions
10. calculate the osmotic pressure of a solution

Module 5 Key Concepts

1. The combined gas law

$$\frac{P_1 V_1}{T_1} = \frac{P_2 V_2}{T_2}$$

This is a combination of Boyle's, Charles's and Gay-Lussac's gas laws. It is used to determine a new temperature, volume, or pressure of a gas given the original temperature, volume and pressure.

2. The ideal gas law

$$PV = nRT$$

This equation is used to calculate the pressure, volume, temperature, or number of moles of a gas given three of the other quantities. It is often used in reaction stoichiometry problems involving gases.

3. Dalton's law of partial pressures

$$P_{Total} = P_1 + P_2 + P_3 + \ldots$$

This equation is used to calculate the total pressure of a gas mixture from the partial pressures of the mixture's component gases.

4. Ion-ion interactions, hydrogen bonding, dipole-dipole interactions, London dispersion forces

These are the four basic intermolecular forces involved in liquids. The strength of these interactions determines the surface tension, vapor pressure, and boiling points of each liquid.

5. Covalent Network Solids, Ionic solids, Metallic solids, Molecular solids

These are the four basic types of solids. The strength of the bonds in solids determines their freezing and boiling points.

---

6. Like Dissolves Like

This rule is a statement of the common phenomenon that polar molecules are readily soluble in other polar molecules and that nonpolar molecules are readily soluble in other nonpolar molecule. However, polar molecules are fairly insoluble in nonpolar molecules.

7. Solute solubility is increased when:

    a. the solvent is *heated* in an *endothermic* dissolution
    b. the solvent is *cooled* in an *exothermic* dissolution
    c. the pressure of a gas (in a liquid) is increased

8. Concentration Units

    a. Percent Concentration

$$\text{Weight/volume (w/v)}\% = \frac{\text{mass of one solution component}}{\text{volume of total solution}} \times 100$$

$$\text{Weight/weight (w/w)}\% = \frac{\text{weight of one solution component}}{\text{weight of total solution}} \times 100$$

$$\text{Volume/volume (v/v)}\% = \frac{\text{volume of one solution component}}{\text{volume of total solution}} \times 100$$

A common set of solution concentration units used in biochemistry

    b. Molarity

$$M = \frac{\text{moles of solute}}{\text{L of solution}}$$

Used in reaction stoichiometry, dilution and osmotic pressure problems

    c. Dilution formula

$$M_1 V_1 = M_2 V_2$$

Used to calculate solution concentrations after dilution

    d. Parts per Million (ppm) or Parts per Billion (ppb)

$$\text{ppm} = \frac{\text{g of solute}}{\text{g of solution}} \times 10^6$$

$$\text{ppb} = \frac{\text{g of solute}}{\text{g of solution}} \times 10^9$$

Used for very dilute solutions especially in biochemistry

9. Freezing-Point Depression of Aqueous Solutions

$$\Delta T_f = \frac{1.86^\circ C}{\text{mol}} \times \text{mol of dissolved particles}$$

This relationship describes how much the freezing temperature of an aqueous solution differs from water's freezing point

10. Osmolarity of solutions

    Osmolarity $= M \times$ **number of particles per formula unit of solute**

This relationship determines the osmolarity of a solution from the solution's molarity.

<u>Sample Exercises</u>
Gas Laws
1. *A sample of a gas initially having a pressure of 1.25 atm and volume of 3.50 L has its volume changed to 7.50 x 10⁴ mL at constant temperature. What is the new pressure of the gas sample?*
   The correct answer is: 0.0583 atm.

$$7.50 \times 10^4 \text{ mL} \left(\frac{1 \text{ L}}{1000 \text{ mL}} \right) = 75.0 \text{ L}$$

$$\frac{P_1 V_1}{T_1} = \frac{P_2 V_2}{T_2} \text{ simplifies to } P_1 V_1 = P_2 V_2 \text{ at constant temperature } (T_1 = T_2)$$

$$1.25 \text{ atm} \times 3.50 \text{ L} = P_2 \times 75.0 \text{ L}$$

$$\frac{1.25 \text{ atm} \times 3.50 \text{ L}}{75.0 \text{ L}} = P_2$$

$$0.0583 \text{ atm} = P_2$$

 CAUTION | It is very important in these problems that the values of P_1 and P_2 or V_1 and V_2 be in the same units. For example, both volumes in this problem must be either in mL or L but not mixed.

2. *A gas sample initially having a pressure of 1.75 atm and a volume of 4.50 L at a 25.0°C is heated to 37.0°C at a pressure of 1.50 atm. What is the gas's new volume?*
   The correct answer is: 5.46 L.

$$\frac{P_1 V_1}{T_1} = \frac{P_2 V_2}{T_2} \text{ where :}$$

$$P_1 = 1.75 \text{ atm}, V_1 = 4.50 \text{ L}, T_1 = 25.0°C = 298.1 \text{ K}$$

$$P_2 = 1.50 \text{ atm and } T_2 = 37.0°C = 310.1 \text{ K}$$

$$V_2 = \frac{P_1 V_1 T_2}{T_1 P_2} = \frac{(1.75 \text{ atm})(4.50 \text{ L})(310.1 \text{ K})}{(298.1 \text{ K})(1.50 \text{ atm})} = 5.46 \text{ L}$$

 CAUTION | All gas law problems involving temperature must be in units of Kelvin. Be absolutely certain that you convert temperatures into Kelvin when working any gas law problems.

3. *A gas sample at a pressure of 3.50 atm and a temperature of 45.0°C has a volume of 1.65 x 10³ mL. How many moles of gas are in this sample?*
   The correct answer is: 0.221 moles.

$$PV = nRT \text{ where}:$$

$$P = 3.50 \text{ atm}, V = 1.65 \times 10^3 \text{ mL} = 1.65 \text{ L}, R = 0.0821 \frac{\text{L atm}}{\text{mol K}}, T = 45.0^\circ\text{C} = 318.1 \text{ K}$$

$$n = \frac{PV}{RT} = \frac{(3.50 \text{ atm})(1.65 \text{ L})}{\left(0.0821 \frac{\text{L atm}}{\text{mol K}}\right)(318.1 \text{ K})} = 0.221 \text{ mol}$$

 TIP | R is the ideal gas constant. In gas laws, its value and units are R = 0.0821 L atm/mol K. This defines the units that we must use in the ideal gas law. P must be in atm, V in L, n in moles, and T in K.

4. **How many grams of $CO_2(g)$ are present in 11.2 L of $CO_2(g)$ at STP?**
    The correct answer is: 22.0 g.

 TIP | STP is a symbol for standard temperature and pressure. When you see those symbols in a problem involving gases, you must assume that the temperature is 273.15 K and the pressure is 1.00 atm or 760 mm Hg.

$$PV = nRT \text{ thus } n = \frac{PV}{RT}$$

$$n = \frac{(1.00 \text{ atm})(11.2 \text{ L})}{\left(0.0821 \frac{\text{L atm}}{\text{mol K}}\right)(273.15 \text{ K})} = 0.500 \text{ mol}$$

$$0.500 \text{ mol}\left(\frac{44.0 \text{ g CO}_2}{1 \text{ mol CO}_2}\right) = 22.0 \text{ g CO}_2$$

5. **A gas mixture composed of Ar, Ne and He has a total pressure of 8.0 atm. If the Ne's partial pressure is 3.0 atm and the He's partial pressure is 1.5 atm, what is the Ar's partial pressure?**
    The correct answer is: 3.5 atm.

$$\mathbf{P_{Total} = P_{Ar} + P_{Ne} + P_{He}}$$

$$\mathbf{P_{Ar} = P_{Total} - (P_{Ne} + P_{He})}$$

$$\mathbf{P_{Ar} = 8.0 \text{ atm} - (3.0 \text{ atm} + 1.5 \text{ atm})}$$

$$\mathbf{P_{Ar} = 3.5 \text{ atm}}$$

Liquids
6. **Arrange these substances by increasing boiling point: CO_2, NaCl, C_2H_5OH, CH_3Cl**
    The correct answer is: $CO_2 < CH_3Cl < C_2H_5OH < NaCl$

Primary Intermolecular Force	Type of Molecule
Ion-ion	Ionic compounds
Hydrogen bonding	Molecules with at least one H atom directly bonded to an O, N, or F atom
Dipole-dipole	Polar molecules
London dispersion forces	Nonpolar molecules

Boiling points for liquids are determined by the strength of the intermolecular forces present in a liquid. In general, the strength of intermolecular forces is: ion-ion interactions > hydrogen bonding > dipole-dipole interactions > London dispersion forces.

The primary intermolecular force between molecules of a given substance is determined by the type of compound and its polarity. Ion-ion interactions are the strongest of these and the intermolecular forces get weaker going down the table. This correlates to a decrease in boiling and/or melting points.

The strongest intermolecular forces in liquid CO_2 (a nonpolar molecule) are London dispersion forces, CH_3Cl (a polar molecule) molecules' strongest intermolecular forces are dipole-dipole interactions, hydrogen bonding is dominant in C_2H_5OH, and NaCl is an ionic compound. Thus the correct order is:

$$CO_2 < CH_3Cl < C_2H_5OH < NaCl$$

Solids

7. ***Arrange these solid substances by increasing melting point: CO_2, KCl, Na, SiO_2***
   The correct answer is: $CO_2 < Na < KCl < SiO_2$

Melting points of solids are determined by the strength of the forces bonding them together. In general, the weakest forces are intermolecular forces found in molecular solids like CO_2, next weakest are metallic bonds as in Na, ionic bonds, such as in KCl, are relatively strong, and the strongest forces are the covalent bonds from atom to atom that bond network covalent species like SiO_2.

The key to melting point problems is determining a solid's classification.
1) **Molecular solids** are always covalent compounds that form individual molecules. Most of the covalent species that you have learned up to now are molecular solids.
2) **Metallic solids** are by far the easiest to classify. Look for a metallic element.
3) **Ionic solids** are the basic ionic compounds that you have learned up to this point.
4) The hardest substances to classify are the **network covalent species**. They are covalent species that form extremely large molecules through extended arrays of atoms that are covalently bonded. Most textbooks have a list of the common network covalent solids which include diamond, graphite, tungsten carbide (WC), and sand (SiO_2). It may be best to memorize these molecules.

8. *How much heat is required to convert 150.0 g of solid Al at 458°C into liquid Al at 758°C? The melting point of Al is 658°C. The specific heats for Al are, C_{solid} = 24.3 J/mol°C and C_{liquid} = 29.3 J/mol°C. The ΔH_{fusion} for Al = 10.6 kJ/mol.* **The correct answer is q = 102.2 kJ.**

△ TIP

This problem involves three separate calculation steps. The final answer is the sum of these three steps. The steps are: 1) heat required to warm the Al from 458°C to its melting point, 658°C, 2) heat required to melt the Al, and 3) heat required to heat the liquid Al from its melting point to the final temperature of 758°C. These steps are illustrated in the diagram below.

◇ CAUTION

Remember, phase changes do NOT have temperature changes associated with them!

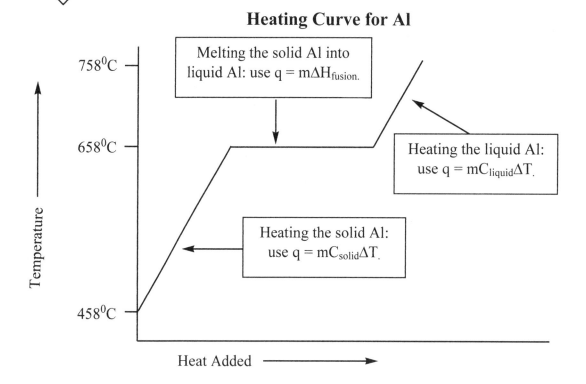

Heating Curve for Al

Melting the solid Al into liquid Al: use q = mΔH_{fusion}.

Heating the liquid Al: use q = m$C_{liquid}\Delta T$.

Heating the solid Al: use q = m$C_{solid}\Delta T$.

Temperature

Heat Added →

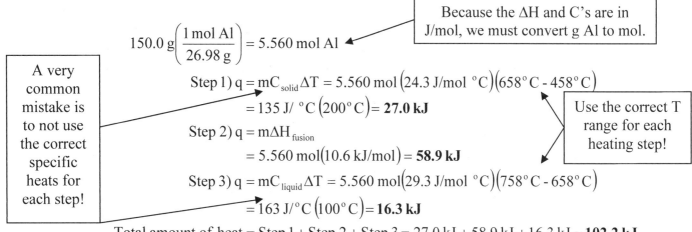

Because the ΔH and C's are in J/mol, we must convert g Al to mol.

$$150.0\ \text{g}\left(\frac{1\ \text{mol Al}}{26.98\ \text{g}}\right) = 5.560\ \text{mol Al}$$

A very common mistake is to not use the correct specific heats for each step!

Step 1) $q = mC_{solid}\Delta T = 5.560\ \text{mol}\left(24.3\ \text{J/mol}\ ^\circ\text{C}\right)\left(658^\circ\text{C} - 458^\circ\text{C}\right)$
$= 135\ \text{J/}^\circ\text{C}\left(200^\circ\text{C}\right) = \mathbf{27.0\ kJ}$

Step 2) $q = m\Delta H_{fusion}$
$= 5.560\ \text{mol}(10.6\ \text{kJ/mol}) = \mathbf{58.9\ kJ}$

Use the correct T range for each heating step!

Step 3) $q = mC_{liquid}\Delta T = 5.560\ \text{mol}\left(29.3\ \text{J/mol}\ ^\circ\text{C}\right)\left(758^\circ\text{C} - 658^\circ\text{C}\right)$
$= 163\ \text{J/}^\circ\text{C}\left(100^\circ\text{C}\right) = \mathbf{16.3\ kJ}$

Total amount of heat = Step 1 + Step 2 + Step 3 = 27.0 kJ + 58.9 kJ + 16.3 kJ = **102.2 kJ**

TIP

Sample exercise 3 involves only one phase change, namely converting solid Al into liquid Al. If a second phase change were included, for example converting solid Al into gaseous Al, the following steps would have to be included.
1) Heating the liquid Al to the boiling point using $q = mC_{liquid}\Delta T$.
2) Boiling the liquid Al using $q = m\Delta H_{vaporization}$.
3) Heating the gaseous Al using $q = mC_{gas}\Delta T$.

9. *Shown below is the phase diagram for water between 1.00 and 760 mm Hg of pressure and -10⁰C to 100⁰C. If the pressure is held constant at 660 mm Hg while the temperature is increased from -10⁰C to 100⁰C, at what temperatures will water melt and boil?*

The correct answers are: $T_{melting} = 0.005^\circ C$ and $T_{boiling} = 90^\circ C$.

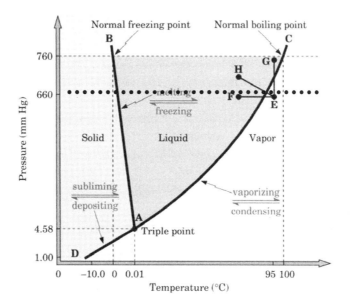

TIP

Notice the horizontal dotted line at 660 mm Hg of pressure stretching from less than -10°C to more than 100 °C. This is a constant pressure (isobar) line. Where this line intersects the melting $\rightleftharpoons$ freezing curve for water will indicate the $T_{melting}$. Likewise the intersection of the dotted line with the vaporizing $\rightleftharpoons$ condensing curve for water indicates the $T_{boiling}$.

Solubility of a Solute in a Given Solvent

10.   Which of the following substances are soluble in water?
$$SiCl_4, NH_3, C_8H_{18}, CaCl_2, CH_3OH, Ca_3(PO_4)_2$$
The correct answer is: only NH_3, $CaCl_2$, and CH_3OH are soluble in water

The "Like Dissolves Like" rule implies that polar species dissolve in polar species and nonpolar species dissolve in nonpolar species. Consequently, nonpolar species do not dissolve in polar species and polar species do not dissolve in nonpolar species. In this example, NH_3 and CH_3OH are both polar covalent compounds so they will dissolve in the highly polar solvent water. $CaCl_2$ is an ionic compound which is water soluble (the solubility rules also apply in these problems). $SiCl_4$ and C_8H_{18} are both nonpolar covalent compounds thus they are insoluble in water. $Ca_3(PO_4)_2$ is an ionic compound that is insoluble in water. Refresh your memory of the solubility rules if necessary.

CAUTION

Keep in mind these two important questions. 1) Are the covalent compounds in the problem polar or nonpolar? 2) Are the ionic compounds in the problem soluble or insoluble based on the solubility rules? (Remember that strong acids and bases are also water soluble.)

Increasing the Solubility of a Solute in a Given Solvent

11.   Given the equation below, which of the following are correct statements?
$$KI(s) \xrightarrow{H_2O} K^+(aq) + I^-(aq) \qquad \Delta H_{dissolution} > 0$$

The correct answer is: only statements a) and f) are true.

a) *Increasing the temperature of the solvent will increase the solubility of the compound in the solvent.*

b) *Decreasing the temperature of the solvent will increase the solubility of the compound in the solvent.*

c) *Changing the temperature of the solvent will not affect the solubility of the compound in the solvent.*

d) *Increasing the pressure of the solute will increase the solubility of the compound in the solvent.*

e) *Increasing the pressure of the solute will decrease the solubility of the compound in the solvent.*

f) *Increasing the pressure of the solute will not affect the solubility of the compound in the solvent.*

There are several important hints in this problem to help you answer it. The $KI_{(s)}$ indicates that this is a solid dissolving in water. Changing the pressure of liquids and solids has no effect on their solubilities. The positive $\Delta H_{dissolution}$ indicates that this is an <u>endothermic</u> process. Heating the solvent for endothermic dissolutions increases the solubility of the solute.

$\Delta H_{dissolution} < 0$ is exothermic. $\Delta H_{dissolution} > 0$ is endothermic.

12. *Given the following dissolution in water equation, which of these changes in conditions are correct statements?*

$$O_2(g) \xrightarrow{\text{H}_2\text{O}} O_2(aq) \quad \Delta H_{dissolution} < 0$$

The correct answer is: only conditions b) and d) are correct

   a) *<u>Increasing</u> the temperature of the solvent will <u>increase</u> the solubility of the compound in the solvent.*
   b) *<u>Decreasing</u> the temperature of the solvent will <u>increase</u> the solubility of the compound in the solvent.*
   c) *Changing the temperature of the solvent will <u>not affect</u> the solubility of the compound in the solvent.*
   d) *<u>Increasing</u> the pressure of the solute will <u>increase</u> the solubility of the compound in the solvent.*
   e) *<u>Increasing</u> the pressure of the solute will <u>decrease</u> the solubility of the compound in the solvent.*
   f) *<u>Increasing</u> the pressure of the solute will <u>not affect</u> the solubility of the compound in the solvent.*

The important hints in this problem are 1) $O_{2(g)}$ indicates that this is a gas dissolving in water. Increasing the pressure of gases has a significant effect on their solubilities. In general, increasing the pressure of a gas will increase its solubility in a liquid. 2) The negative $\Delta H_{dissolution}$ indicates that this is an <u>exothermic</u> process. Heating the solvent for exothermic dissolutions decreases the solubility of the solute. Cooling the solvent increases the solubility of the solute in exothermic dissolutions.

1) Pay attention to whether the substance being dissolved is a solid, liquid, or gas. That will tell you if the changing pressure condition is applicable. 2) Pay attention to whether or not the dissolution is endo- or exothermic. That is your hint as to heating or cooling the solvent will increase the solubility of the substance. Both of these effects are ramifications of LeChatelier's principle.

Concentration Units
13. *An aqueous sulfuric acid solution that is 30.0 % w/w has a density of 1.225 g/mL. What is the concentration of this solution in w/v%?*
   The correct answer is 36.8 % w/v.

Remember that 30.0% w/w $H_2SO_4 = \dfrac{30.0 \text{ g of } H_2SO_4}{100.0 \text{ g of solution}}$. 30.0% w/w indicates the mass of the solution, not the volume which is needed to determine the % w/v. The solution's density will help us calculate the solution's volume. To make the calculation simple, we can assume that we have 100.0 g of this 30.0% w/w solution.

$$100.0 \text{ g solution}\left(\dfrac{1.00 \text{ mL}}{1.225 \text{ g}}\right) = 81.63 \text{ mL of } 30.0\% \text{ w/w } H_2SO_4$$

density of the solution

volume of the solution

$$\% \text{ w/v} = \left(\dfrac{30.0 \text{ g of } H_2SO_4}{81.63 \text{ mL of solution}}\right) = 36.8 \% \text{ w/v } H_2SO_4$$

CAUTION

Solution concentration units indicate the number of solute particles contained in an amount of the solution. In these % concentration units you must pay careful attention to the units necessary for each one of the concentration units. Notice that in the problem above we have converted the denominator from mass (g) to volume (v).

14. *An aqueous sucrose, $C_{12}H_{22}O_{11}$, solution contains 11.0 g of $C_{12}H_{22}O_{11}$ dissolved in 89.0 g of water and has a density of 1.0432 g/mL. What is the concentration of this solution in molarity (M)?*
The correct answer is 0.335 M.

The total mass of this solution is 11.0 g + 89.0 g = 100.0 g of solution. From the mass and density of the solution we can determine the volume of the solution.

density of the solution

$$100.0 \text{ g of solution}\left(\dfrac{1.000 \text{ mL}}{1.0432 \text{ g}}\right) = 95.86 \text{ mL}\left(\dfrac{1.00 \text{ L}}{1000 \text{ mL}}\right) = 0.09586 \text{ L}$$

To determine the solution's molarity we must also know the number of moles of sucrose.

$$11.0 \text{ g of } C_{12}H_{22}O_{11}\left(\dfrac{1 \text{ mole of } C_{12}H_{22}O_{11}}{342.3 \text{ g of } C_{12}H_{22}O_{11}}\right) = 0.0321 \text{ moles of } C_{12}H_{22}O_{11}$$

Proper combination of the moles of sucrose and the solution's volume yields the solution's molarity.

$$M = \dfrac{\text{moles of sucrose}}{\text{L of solution}} = \dfrac{0.0321 \text{ moles of sucrose}}{0.0959 \text{ L of solution}} = \textbf{0.335 M}$$

15. *An aqueous solution contains 2.5×10^{-5} g of Cs dissolved in 1.00 kg of water. What is the concentration of this solution in ppm and ppb?*
The correct answer is 0.025 ppm and 25 ppb.

$$1.00 \text{ kg of } H_2O = 1000 \text{ g of } H_2O$$

$$\dfrac{2.5 \times 10^{-5} \text{ g of Cs}}{1000 \text{ g of } H_2O} = 2.5 \times 10^{-8}$$

$$\left(2.5 \times 10^{-8}\right) \times 10^{6} = 0.025 \text{ ppm}$$

$$\left(2.5 \times 10^{-8}\right) \times 10^{9} = 25 \text{ ppb}$$

Solution Dilution

16. We need to make 250.0 mL of 0.375 M aqueous NaOH solution from a 6.00 M NaOH(aq) solution. How much of the 6.00 M solution is required to make the 0.250 M solution?

The correct answer is 15.6 mL of the 6.00 *M* solution are required.

The dilution formula, $M_1V_1 = M_2V_2$, is used to determine the amounts of concentrated solution necessary to make a dilute solution. The subscript ones indicate the quantities for the concentrated solution while the two's refer to the dilute solution. In this problem we are given the concentrations of both the concentrated and dilute solutions as well as the volume of the dilute solution. Thus we know M_1, M_2, and V_2. All we have to do is solve for V_1.

$$M_1V_1 = M_2V_2 \therefore V_1 = \frac{M_2V_2}{M_1}$$

$$V_1 = \frac{0.375M \times 250.0 \text{ mL}}{6.00M}$$

$$V_1 = 15.6 \text{ mL}$$

Freezing Point Depression of Aqueous Solutions

17. If 11.0 g of sucrose, a nonelectrolyte, are dissolved in 89.0 g of water, at what temperature will this solution freeze at 1.00 atm of pressure?

The correct answer is -0.0597°C.

From the freezing point depression formula we realize that we must determine the number of moles of sucrose in 11.0 g.

$$11.0 \text{ g of C}_{12}\text{H}_{22}\text{O}_{11} \left(\frac{1 \text{ mole of C}_{12}\text{H}_{22}\text{O}_{11}}{342.3 \text{ g of C}_{12}\text{H}_{22}\text{O}_{11}} \right) = 0.0321 \text{ moles of C}_{12}\text{H}_{22}\text{O}_{11}$$

Substitution of the number of moles of sucrose into the freezing point depression formula yields the amount the freezing point will go down.

$$\Delta T_f = \frac{1.86°C}{\text{mol}} \times \text{mol of dissolved particles}$$

$$\Delta T_f = \frac{1.86°C}{\text{mol}} \times 0.0321 \text{ mol} = 0.0597°C$$

Addition of the sugar causes the freezing point of the solution to be less (depressed) than that of pure water. Thus the final freezing point of this solution is pure water's freezing point minus the ΔT_f.

$$\text{Solution's freezing point} = 0.0000°C - 0.0597°C = -0.0597°C$$

18. If 11.0 g of sodium chloride, an electrolyte, are dissolved in 89.0 g of water, at what temperature will this solution freeze at 1.00 atm of pressure?

The correct answer is -0.699°C.

Just as in the previous problem, we must determine the number of moles of solute in the solution. The 11.0 g of NaCl are the solvent.

$$11.0 \text{ g of NaCl}\left(\frac{1 \text{ mole of NaCl}}{58.44 \text{ g}}\right) = 0.188 \text{ moles of NaCl}$$

However, because NaCl is a strong electrolyte when 1 mole of NaCl is dissolved in solution it forms two moles of dissolved species, 1 mole of Na^+ ions and 1 mole of Cl^- ions. Consequently, we must multiply the moles of NaCl in this solution by two to determine the moles of dissolved particles.

moles of dissolved particles in NaCl solution $= 2 \times 0.188 \text{ moles} = 0.376 \text{ moles}$

Substitution of the number of moles of NaCl into the freezing point depression formula yields the amount the freezing point will go down.

$$\Delta T_f = \frac{1.86^\circ C}{mol} \times \text{mol of dissolved particles}$$

$$\Delta T_f = \frac{1.86^\circ C}{mol} \times 0.376 \text{ mol} = 0.699^\circ C$$

Then we determine the final freezing point of this solution as before.

Solution's freezing point $= 0.000^\circ C - 0.699^\circ C = -0.699^\circ C$

Osmolarity of Solutions

19. If 11.0 g of sodium chloride, an electrolyte, are dissolved in sufficient water to make 100 mL (0.100 L) of solution, what is the osmolarity of the solution? Is the solution hyper- or hypotonic? An isotonic NaCl solution has an osmolarity of 0.30.

The correct answer is 3.76 osmolar and the solution is hypertonic.

From the previous problem we know that 11.0 g of NaCl is 0.188 moles of NaCl. Thus the solution's molarity is $\frac{0.188 \text{ mol}}{0.100 \text{ L}} = 1.88\ M$. To determine the osmolarity we must take into account that NaCl dissolves into two particles, Na^+ ions and Cl^- ions. Thus we can determine the osmolarity of the solution.

Osmolarity $= 1.88\ M \times 2 = 3.76 \text{ osmolar}$

Since isotonic NaCl solutions are 0.30 osmolar and this solution is 3.76 osmolar, this solution is 12.5 times more concentrated than an isotonic NaCl solution and thus is hypertonic.

Module 6
Chemical Kinetics and Equilibrium

Introduction

Chemical kinetics describes how quickly chemical reactions occur. There are several factors that chemists use to change a reaction's rate including temperature and the concentrations of the reactants. This module describes:

1. how to determine the rate of a reaction
2. collision and transition state theories of reactions
3. methods to change a reaction's rate
4. concept of equilibrium
5. equilibrium constants
6. Le Châtelier's principle

Module 6 Key Concepts

1. $$\text{reactant reaction rate} = \frac{-\text{ amount of reactant consumed}}{\text{amount of time}}$$

   $$\text{product reaction rate} = \frac{+\text{ amount of product formed}}{\text{amount of time}}$$

   Chemical reaction rates are determined by the amounts of the reactants consumed or products formed in an amount of time. Notice that the reactant amount decreases with time, indicated by the $-$ sign, while the product amount increases with time, requiring a $+$ sign.

2. **For chemical reactions to occur the reacting molecules must meet the following conditions :**
   a. molecules must collide
   b. colliding molecules must have sufficient energy to break reactant chemical bonds and from product chemical bonds
   c. molecules must be properly oriented
   d. a measure of the energy required to break and reform bonds is indicated by the reaction's activation energy

3. **Reaction rates depend upon four factors**
   a. nature of the reactants
   b. concentrations of the reactants
   c. temperature of the reactants
   d. presence of a catalyst

4. **For the reaction $aA + bB \rightleftharpoons cC + dD$ where a, b, c, and d are the stoichiometric coefficients for the reaction, the equilibrium constant,**

   $$K = \frac{[C]^c [D]^d}{[A]^a [B]^b}$$

   Equilibrium concentrations must be used in K.

Sample Exercises
Reaction Rates
1. *The initial H_2 concentration for this reaction at 15^oC is 16.8 M. After the reaction has proceeded for 4.00 seconds the H_2 concentration is 5.60 M. What is this reaction's rate of reaction?*

$$H_2(g) \ + \ Cl_2(g) \ \rightarrow \ 2\,HCl(g)$$

The negative sign indicates that the H_2 concentration is decreasing with time.

The correct answer is - 2.80 M/s.

$$\text{rate of reaction} = -\left[\frac{16.8\ M\text{ - }5.60\ M}{4.00\ s}\right] = -\left[\frac{11.2\ M}{4.00\ s}\right] = -2.8\ M/s$$

2. *For this reaction*

$$H_2(g) \ + \ Cl_2(g) \ \rightarrow \ 2\,HCl(g)$$
the reaction rate doubles for every 10^oC increase in temperature. What is the reaction rate at 45^oC?

The correct answer is - 22.4 M/s.

Multiply the factors of 2, **do not add them**. Rate of -2.80 M/s is from Sample Exercise 1.

Determine the temperature change and number of 10.0^oC increments.

Temperature rise = 45.0^oC - 15.0^oC = 30.0^oC

Number of 10.0^oC rises = $30.0^oC/10.0^oC$ = 3

3 doublings of the rate = $2 \times 2 \times 2 = 8$ times faster

$-2.80\ M/s \times 8 = -22.4\ M/s$

Collision Theory
3. *Choose the statements which describe necessary conditions for reacting molecules.*
   a. *A Cl_2 molecule slowly colliding with an H_2 molecule.*
   b. *A Cl_2 molecule rapidly sliding by an H_2 molecule without collision.*
   c. *A Cl_2 molecule quickly colliding with an H_2 molecule.*
   d. *A Cl_2 molecule colliding with an H_2 molecule in this fashion.*

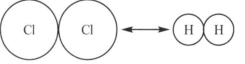

   e. *A Cl_2 molecule colliding with an H_2 molecule in this fashion.*

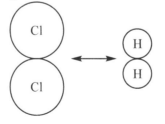

The correct answer is statements c. and e.

For a molecular collision to be effective in converting reactant molecules into product molecules the collision must be of sufficient energy, as indicated in this image,

and the molecules must be properly oriented so that the reacting atoms can be exchanged, as indicated in this image for the reaction of H_2O and HCl.

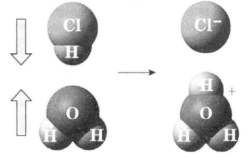

4. *Increasing which one of the indicated energies in this energy diagram will decrease the rate of the reaction of H_2O with HCl?*
    a.   b.   c.   or   d.

The correct answer is b.

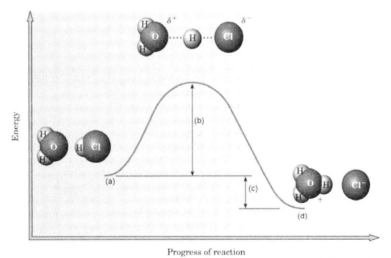

Answer (b) indicates the reaction's activation energy. If the activation energy is increased, the reaction rate decreases.

5. *Choose the rate factor which affects the activation energy.*
    a. nature of reactants
    b. temperature
    c. catalyst
    d. reactant concentrations

The correct answer is c.

Catalysts are the only rate factor which can change the activation energy. By lowering the activation energy, catalysts can make reactions occur many times faster than possible without the catalyst.

CAUTION

It is a good idea to understand the effects of each of the four rate factors used in Sample Exercise 3. Frequently, test questions are designed to assess your understanding of these factors. For example, you should know if increasing temperature or a reactant's concentration will make the reaction rate increase or decrease.

Rate Constants

6. *Determine the value of the rate constant for the reaction*

$$NH_2NO_2(aq) \rightarrow N_2O(g) + H_2O(\ell)$$

if the rate law for this reaction is rate = - k[NH₂NO₂], the rate is 1.86 × 10⁻⁴ M/s and the starting [NH₂NO₂] = 2.00 M.

The correct answer is $-9.30 \times 10^{-5}\ ^1\!/_s$.

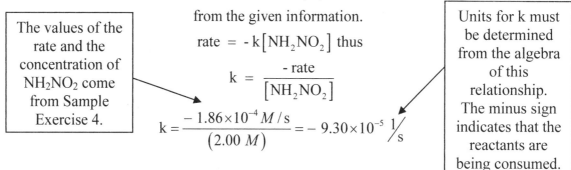

The values of the rate and the concentration of NH₂NO₂ come from Sample Exercise 4.

The value of the rate constant, k, can be determined from the given information.

$$rate = -k\left[NH_2NO_2\right] \text{ thus}$$

$$k = \frac{-rate}{\left[NH_2NO_2\right]}$$

$$k = \frac{-1.86 \times 10^{-4}\,M/s}{(2.00\,M)} = -9.30 \times 10^{-5}\ ^1\!/_s$$

Units for k must be determined from the algebra of this relationship. The minus sign indicates that the reactants are being consumed.

7. *The following reaction*

$$NH_2NO_2(aq) \rightarrow N_2O(g) + H_2O(\ell)$$

obeys this rate law: rate = k[NH₂NO₂] with the rate constant, k, having a value of $9.30 \times 10^{-5}\ ^1\!/_s$. If the initial [NH₂NO₂] = 6.00 M, what is the reaction rate?

The correct answer is $5.58 \times 10^{-4}\ M\!/_s$.

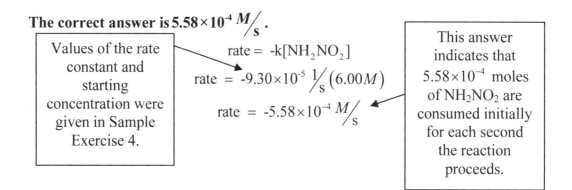

Values of the rate constant and starting concentration were given in Sample Exercise 4.

$$rate = -k[NH_2NO_2]$$

$$rate = -9.30 \times 10^{-5}\ ^1\!/_s\,(6.00M)$$

$$rate = -5.58 \times 10^{-4}\ M\!/_s$$

This answer indicates that 5.58×10^{-4} moles of NH₂NO₂ are consumed initially for each second the reaction proceeds.

Form of the Equilibrium Constant

8. *Write the equilibrium constant expression for this reaction.*

$$H_2(g) + I_2(g) \rightleftharpoons 2\,HI(g)$$

The correct answer is $K = \dfrac{[HI]^2}{[H_2][I_2]}$.

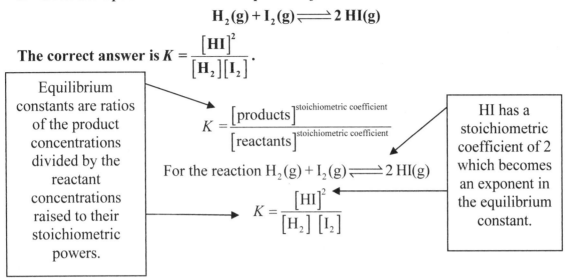

Equilibrium constants are ratios of the product concentrations divided by the reactant concentrations raised to their stoichiometric powers.

$$K = \frac{[products]^{\text{stoichiometric coefficient}}}{[reactants]^{\text{stoichiometric coefficient}}}$$

For the reaction $H_2(g) + I_2(g) \rightleftharpoons 2\,HI(g)$

$$K = \frac{[HI]^2}{[H_2]\,[I_2]}$$

HI has a stoichiometric coefficient of 2 which becomes an exponent in the equilibrium constant.

Value of the Equilibrium Constant

9. *For the following reaction at 298 K, the equilibrium concentrations are [H₂] = 1.50 M, [I₂] = 2.00 M, and [HI] = 3.46 M. What is the value of the equilibrium constant, Kc, for this reaction at 298 K?*

$$H_2(g) + I_2(g) \rightleftharpoons 2\,HI(g)$$

The correct answer is: 4.00.

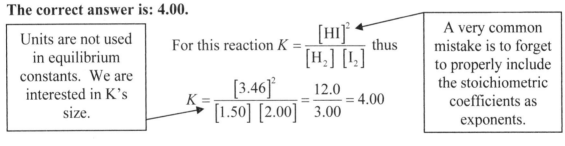

Units are not used in equilibrium constants. We are interested in K's size.

For this reaction $K = \dfrac{[HI]^2}{[H_2]\,[I_2]}$ thus

$$K = \frac{[3.46]^2}{[1.50]\,[2.00]} = \frac{12.0}{3.00} = 4.00$$

A very common mistake is to forget to properly include the stoichiometric coefficients as exponents.

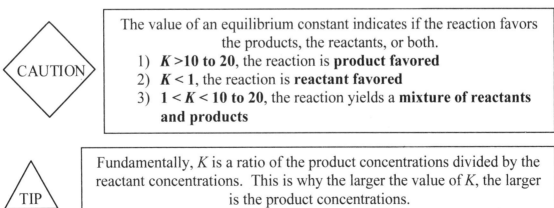

CAUTION

The value of an equilibrium constant indicates if the reaction favors the products, the reactants, or both.
1) $K > 10$ to 20, the reaction is **product favored**
2) $K < 1$, the reaction is **reactant favored**
3) $1 < K < 10$ to 20, the reaction yields a **mixture of reactants and products**

TIP

Fundamentally, K is a ratio of the product concentrations divided by the reactant concentrations. This is why the larger the value of K, the larger is the product concentrations.

Le Châtelier's Principle

10. What is the effect (increased relative product concentration, increased relative reactant concentration, no effect) of each of these changes on the position of equilibrium of this reaction at 298 K?

$$2\ NO_2(g) \rightleftharpoons N_2O_4(g) \quad \Delta H^0_{rxn} = -57.2\ kJ/mol$$

a) Increasing the temperature of the reaction
b) Removing some NO_2 from the reaction vessel.
c) Adding some N_2O_4 to the reaction vessel.
d) Increasing the pressure in the reaction vessel by adding an inert gas.
e) Decreasing by half the size of the reaction vessel.
f) Introducing a catalyst into the reaction vessel.

The correct answers are a) increasing the reaction temperature will increase the relative reactant concentration b) removing some NO_2 will increase the relative reactant concentration c) adding some N_2O_4 will increase the relative reactant concentration d) adding an inert gas will have no effect on either the reactant or product concentrations e) decreasing the reaction vessel size by half will increase the relative product concentration f) introducing a catalyst will have no effect on either the reactant or product concentrations.

All of these changes are illustrations of Le Châtelier's principle: If an external stress is applied to a system at equilibrium is stressed, the system responds in some fashion to partially relieve that stress.

a) For exothermic reactions: increasing the temperature increases the reactant concentrations at the expense of the product concentrations, decreasing the temperature increases the product concentrations at the expense of the reactant concentrations. Endothermic reactions behave oppositely. In this exercise the negative ΔH^0_{rxn} indicates that the reaction is exothermic, thus increasing the temperature increases the relative reactant concentrations.

b) If a reactant's concentration is decreased below the equilibrium concentration, the position of equilibrium will change to restore concentrations that correspond to those predicted by the equilibrium constant. In this exercise, removing some NO_2 from the reaction vessel decreases the $[NO_2]$. The reaction equilibrium responds to this stress by increasing the $[NO_2]$ and decreasing the $[N_2O_4]$, an equilibrium position shift to the reactant side. Adding NO_2 would cause the position of equilibrium to shift to the product side.

c) *If a product's concentration is increased above the equilibrium concentration, the equilibrium position will shift to restore concentrations of products and reactants that correspond to those predicted by the equilibrium constant.* Adding some N_2O_4 to the reaction vessel increases the $[N_2O_4]$ above the equilibrium concentration. The reaction equilibrium responds to this stress by decreasing the $[N_2O_4]$ and increasing the $[NO_2]$, an equilibrium position shift to the reactant side. Removing N_2O_4 would shift the position of equilibrium to the product side.

d) *Adding an inert gas to the reaction mixture has no effect on the equilibrium position because the concentrations of the gases are not changed.* This is a common misconception for students.

e) *If the volume of the reaction vessel is changed, the concentrations of the gases are changed because for gases, M ∝ n/V. If the vessel's volume is decreased, the equilibrium position will shift to the side that has the fewest moles of gas.* In this exercise the right or product side has the fewest moles.

f) *Adding a catalyst has no effect on the position of equilibrium.* Catalysts change the rates of reactions but not positions of equilibrium.

Module 7
Acids and Bases

Introduction

Two common acid-base theories help explain many organic and biochemical reactions: the Arrhenius and the Brønsted-Lowry theories. In aqueous solutions, acid and base strengths can be measured using pH, pOH, and acid-base titrations and these strengths are reflected in their ionization constants. Weak acids and bases can be combined with their salts to form buffer solutions. This module will help you understand:

1. the distinctions and the commonalities between the Arrhenius and the Brønsted-Lowry theories
2. how to use the concepts of pH and pOH
3. how to use acid ionization constants
4. how to determine acid or base concentrations using titrations
5. how buffers are made
6. how to calculate a buffer's pH.

Module 7 Key Concepts

1. **Arrhenius Acid-Base Theory**

   Acid: produces hydronium ions (H_3O^+) in aqueous solutions
   Base: produces OH^- in aqueous solutions
   Arrhenius theory applies only to compounds containing either acidic H^+ or basic OH^- and can increase their concentrations in aqueous solutions.

2. **Brønsted-Lowry Acid-Base Theory**

   Acid: proton donor
   Base: proton acceptor
   This theory is less restrictive. Bases do not have to contain OH^-, and the compounds do not have to be in aqueous solution.

3. **For the weak acid equilibrium** $HA \rightleftharpoons H^+ + A^-$ $K_a = \dfrac{\left[H^+\right]\left[A^-\right]}{\left[HA\right]}$

   K_a is the weak acid ionization constant. K_a values indicate the relative strengths of acids species in aqueous solutions. K_a values are tabulated in an appendix in your textbook.

4. **The ionization constant for water,** $K_w = \left[H_3O^+\right]\left[OH^-\right] = 1.00 \times 10^{-14}$
   $$14 = pH + pOH$$

   K_w is used to calculate either the hydronium or hydroxide ion concentration in aqueous solutions given the concentration of either one of these ions. A logarithmic form of the first equation, given on the 2nd line, relates the solutions' pH and pOH.

5.

$$pH = -\log\left[H^+\right]$$

$$pOH = -\log\left[OH^-\right]$$

These equations define pH and pOH. pH is a condensed method to write the H^+ or H_3O^+ concentration in aqueous solutions. pOH is an equivalent method of writing the aqueous OH^- concentration.

6. **The Henderson-Hasselbalch equations:**

For an acidic buffer solution $pH = pK_a + \log\dfrac{\left[A^-\right]}{\left[HA\right]}$

For a basic buffer solution $pOH = pK_b + \log\dfrac{\left[B^+\right]}{\left[BOH\right]}$

These are used to find the pH of buffer solutions given concentrations of the salt and acid for acidic buffers, or the salt and base for basic buffers.

Sample Exercises
Arrhenius Acid-Base Theory
1. Which of these compounds are Arrhenius acids and which are Arrhenius bases?
  HCl, NaOH, H_2SO_4, BCl_3, Na_2CO_3, $Ba(OH)_2$, C_2H_4

The correct answer is HCl and H_2SO_4 are Arrhenius acids and NaOH and $Ba(OH)_2$ are Arrhenius bases. Na_2CO_3, BF_3 and C_2H_4 are neither Arrhenius acids nor bases.

To identify Arrhenius acids look for compounds that dissociate or ionize in water forming H_3O^+. To identify Arrhenius bases look for compounds that dissociate or ionize in water producing OH^-.

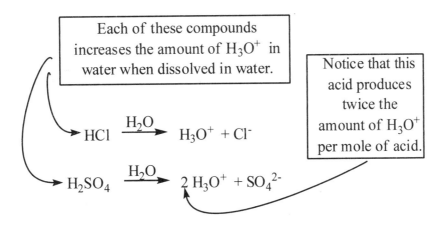

Each of these compounds increases the amount of H_3O^+ in water when dissolved in water.

Notice that this acid produces twice the amount of H_3O^+ per mole of acid.

$$HCl \xrightarrow{H_2O} H_3O^+ + Cl^-$$

$$H_2SO_4 \xrightarrow{H_2O} 2\,H_3O^+ + SO_4{}^{2-}$$

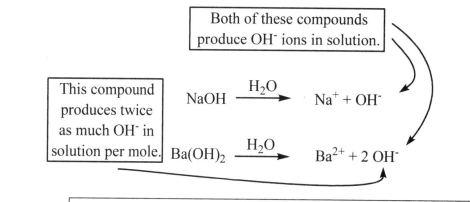

Both of these compounds produce OH⁻ ions in solution.

This compound produces twice as much OH⁻ in solution per mole.

$$NaOH \xrightarrow{H_2O} Na^+ + OH^-$$

$$Ba(OH)_2 \xrightarrow{H_2O} Ba^{2+} + 2\,OH^-$$

CAUTION

It is relatively easy to see that BF_3 and Na_2CO_3 are not acids or bases under this theory since they do not contain H or OH. C_2H_4 may be a little trickier. It does contain H atoms. Do not let this confuse you! The H atoms are not acidic in this case because the C-H bond is too strong to be easily broken.

Brønsted-Lowry Acid-Base Theory

2. Which of these compounds can be classified as Brønsted-Lowry acids and bases?
HF, NH₃, H₂SO₄, BCl₃, Na₂CO₃, K₂S

The correct answer is: HF and H_2SO_4 are Brønsted-Lowry acids; NH_3 and Na_2CO_3 are Brønsted-Lowry bases. BCl_3 and K_2S are neither.

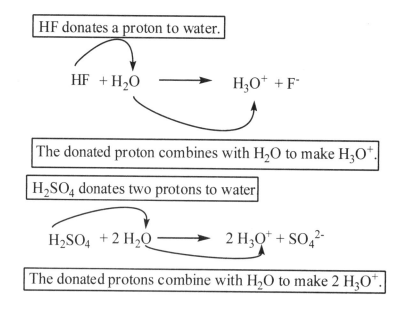

HF donates a proton to water.

$$HF + H_2O \longrightarrow H_3O^+ + F^-$$

The donated proton combines with H_2O to make H_3O^+.

H_2SO_4 donates two protons to water

$$H_2SO_4 + 2\,H_2O \longrightarrow 2\,H_3O^+ + SO_4^{2-}$$

The donated protons combine with H_2O to make $2\,H_3O^+$.

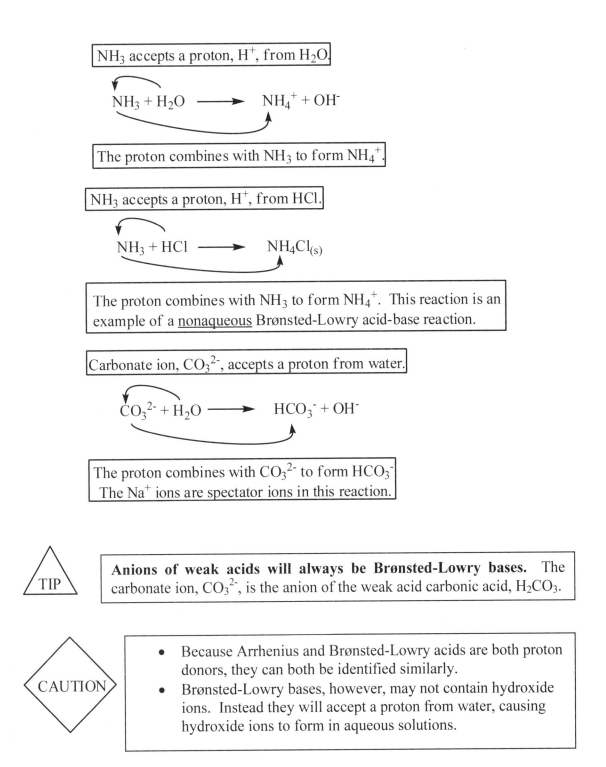

NH$_3$ accepts a proton, H$^+$, from H$_2$O.

$$NH_3 + H_2O \longrightarrow NH_4^+ + OH^-$$

The proton combines with NH$_3$ to form NH$_4^+$.

NH$_3$ accepts a proton, H$^+$, from HCl.

$$NH_3 + HCl \longrightarrow NH_4Cl_{(s)}$$

The proton combines with NH$_3$ to form NH$_4^+$. This reaction is an example of a nonaqueous Brønsted-Lowry acid-base reaction.

Carbonate ion, CO$_3^{2-}$, accepts a proton from water.

$$CO_3^{2-} + H_2O \longrightarrow HCO_3^- + OH^-$$

The proton combines with CO$_3^{2-}$ to form HCO$_3^-$. The Na$^+$ ions are spectator ions in this reaction.

Conjugate acid-base pairs: In Brønsted-Lowry theory, two species that differ by the presence or absence of a proton are conjugate acid-base pairs. Each Brønsted-Lowry acid has a conjugate base (the base does not have the H$^+$), and each Brønsted-Lowry base has a conjugate acid (which has the H$^+$).

3. Identify the Brønsted-Lowry acid-base conjugate pairs in these reactions.

$$CH_3COOH + H_2O \rightleftharpoons CH_3COO^- + H_3O^+$$

$$F^- + H_2O \rightleftharpoons HF + OH^-$$

The correct answer is CH₃COOH is an acid; CH₃COO⁻ is its conjugate base
H₂O is a base; H₃O⁺ is its conjugate acid
F⁻ is a base; HF is its conjugate acid
H₂O is an acid; OH⁻ is its conjugate base

$$CH_3COOH + H_2O \rightleftharpoons CH_3COO^- + H_3O^+$$

> CH₃COOH donates a proton to H₂O making it an acid. It's conjugate base, CH₃COO⁻, differs from the acid by the loss of a single proton, H⁺.

$$CH_3COOH + H_2O \rightleftharpoons CH_3COO^- + H_3O^+$$

> H₂O accepts a proton from CH₃COOH making it a base. It's conjugate acid, H₃O⁺, differs from the base by the addition of a single proton, H⁺.

To identify acid-base conjugate pairs, look for two compounds that differ by the presence or absence of an H⁺. If the compound on the reactant side contains the H⁺ and on the product side has lost the H⁺, the reactant is an acid and the product is its conjugate base.

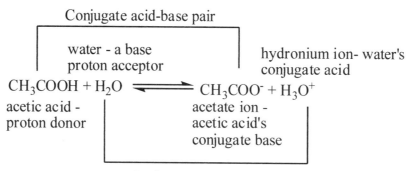

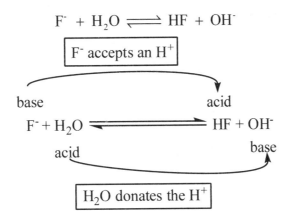

$$F^- + H_2O \rightleftharpoons HF + OH^-$$

F^- accepts an H^+

base   acid
$$F^- + H_2O \rightleftharpoons HF + OH^-$$
acid   base

H_2O donates the H^+

CAUTION

Compare the two reactions used as examples above. In one reaction H_2O is a base and in the other it is an acid. Species that can be either acidic or basic are called *amphoteric*. **In Brønsted-Lowry theory all acid-base reactions are a competition between stronger and weaker acids or bases.** In the CH_3COOH reaction, the stronger acid is CH_3COOH, thus water acts as a base in its presence. In the F^- reaction, H_2O is the stronger acid behaving as an acid in this reaction. Water can be either an acid or a base in the presence of a stronger acid or base.

TIP

List of Common Inorganic Strong Acids and Bases

Strong Acids

HCl   hydrochloric acid
HBr   hydrobromic acid
HI   hydroiodic acid
HNO_3 nitric acid
H_2SO_4 sulfuric acid
$HClO_4$ perchloric acid

Strong Bases

LiOH   lithium hydroxide
NaOH   sodium hydroxide
KOH   potassium hydroxide
$Ba(OH)_2$ barium hydroxide

It is a very good idea for you to be familiar with this list of strong acids and bases.

K_a and pK_a Values

4. *Given below are the K_a values for three acids. Arrange the acids in order of increasing acid strength. K_a for $H_2CO_3 = 4.3 \times 10^{-7}$. K_a for $HCl = \infty$. K_a for $H_3PO_4 = 7.5 \times 10^{-3}$?*
The correct answer is $HCl > H_3PO_4 > H_2CO_3$.

K_a values and their logarithmic form, pK_a values, are reliable indictors of an acid's strength. Large K_a values indicate strong acids. Small K_a values are associated with weak acids. HCl has the largest K_a value, ∞ (infinity) in this trio of acids and is the strongest acid. (Look at the previous insight, HCl is one of the strong acids all of which have a K_a value of ∞. H_3PO_4, the second strongest acid, has the second largest K_a value. H_2CO_3 is the weakest acid with the smallest K_a value. pK_a values, because they are logarithmic, have the opposite trend. The larger the pK_a value, the weaker the acid.

5. *Arrange the following species in order of increasing base strength:. HCO_3^-, Cl^-, $H_2PO_4^-$?*

The correct answer is $Cl^- < H_2PO_4^- < HCO_3^-$

The best approach to a problem like this is to recognize the acid or base the species are derived from. (Do this by adding H^+ to the species and determining their conjugate acids.)

The Cl^- ion is derived from HCl, $H_2PO_4^-$ derives from H_3PO_4, and finally HCO_3^- derives from H_2CO_3.

Now we can easily compare the conjugate acid strengths and determine their base strengths.
-Comparing acid strengths, HCl is the strongest acid, H_3PO_4 is the next strongest acid, and H_2CO_3 is the weakest acid.
-Since the Cl^- ion is the conjugate base of the strong acid HCl, it is the weakest base. The $H_2PO_4^-$ ion is the conjugate base of the weak acid H_3PO_4, thus it is a stronger base than Cl^-. Finally, the HCO_3^- ion is the conjugate base of the very weak acid H_2CO_3 making it the strongest base.

Some important things to remember about acids and bases are:
1) **The stronger the acid, the weaker the conjugate base.**
2) **The weaker the acid, the stronger the conjugate base.**
3) **The stronger the base, the weaker the conjugate acid.**
4) **The weaker the base, the stronger the conjugate acid.**

pH, pOH, and K_w
6. *What is the [OH⁻] in an aqueous solution with a pH of 5.25?*
   The correct answer is: [OH⁻] = 1.8 x 10⁻⁹ M.

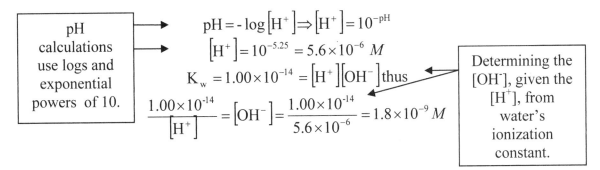

$$pH = -\log\left[H^+\right] \Rightarrow \left[H^+\right] = 10^{-pH}$$

$$\left[H^+\right] = 10^{-5.25} = 5.6 \times 10^{-6}\ M$$

$$K_w = 1.00 \times 10^{-14} = \left[H^+\right]\left[OH^-\right]\ \text{thus}$$

$$\frac{1.00 \times 10^{-14}}{\left[H^+\right]} = \left[OH^-\right] = \frac{1.00 \times 10^{-14}}{5.6 \times 10^{-6}} = 1.8 \times 10^{-9}\ M$$

pH calculations use logs and exponential powers of 10.

Determining the [OH⁻], given the [H⁺], from water's ionization constant.

7. What is the pH of an aqueous solution that has a [OH] = 3.45 x 10⁻³?
   The correct answer is: pH = 11.538.

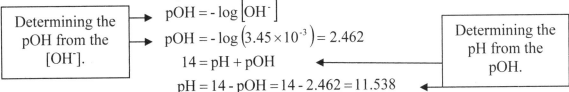

Determining the pOH from the [OH⁻].

$$pOH = -\log\left[OH^-\right]$$

$$pOH = -\log\left(3.45 \times 10^{-3}\right) = 2.462$$

$$14 = pH + pOH$$

$$pH = 14 - pOH = 14 - 2.462 = 11.538$$

Determining the pH from the pOH.

Solution Concentration from a Titration

8. How many mL of 0.125 M HCl are required to exactly neutralize 25.0 mL of an aqueous 0.025 M Sr(OH)₂ solution?
   The correct answer is: 50.0 mL.

 TIP

The word "neutralize" is your clue that this is a titration problem. In this case, you should also note that it is the reaction of a strong acid with the dihydroxy strong base, Sr(OH)₂. **In all titrations, you must write a balanced chemical reaction before doing any calculations**.

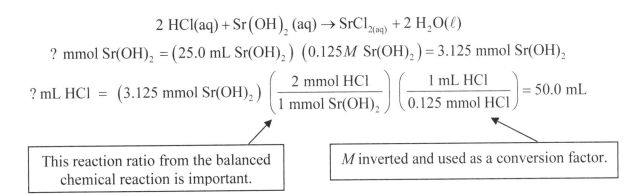

$$2\ HCl(aq) + Sr\left(OH\right)_2 (aq) \rightarrow SrCl_{2(aq)} + 2\ H_2O(\ell)$$

$$?\ \text{mmol } Sr(OH)_2 = \left(25.0\ \text{mL } Sr(OH)_2\right)\left(0.125 M\ Sr(OH)_2\right) = 3.125\ \text{mmol } Sr(OH)_2$$

$$?\ \text{mL HCl} = \left(3.125\ \text{mmol } Sr(OH)_2\right)\left(\frac{2\ \text{mmol HCl}}{1\ \text{mmol } Sr(OH)_2}\right)\left(\frac{1\ \text{mL HCl}}{0.125\ \text{mmol HCl}}\right) = 50.0\ \text{mL}$$

This reaction ratio from the balanced chemical reaction is important.

M inverted and used as a conversion factor.

Buffer Solutions

9. What is the pH of a solution that is 0.100 M in acetic acid, CH₃COOH, and 0.025 M in sodium acetate, NaCH₃COO? For acetic acid Kₐ = 1.8 × 10⁻⁵.
   The correct answer is pH = 4.14.

Key clues to recognizing a buffer problem:
  a) The solution will contain a soluble salt dissolved in either a weak acid or a weak base. Concentrations or amounts of both will be present in the problem.
  b) The salt must be the conjugate partner of the weak acid or weak base. In this sample exercise $NaCH_3COO$ is the soluble salt of the of acetic acid, CH_3COOH.
  c) Henderson-Hasselbalch equations are a simple method to find the buffer solution's pH.

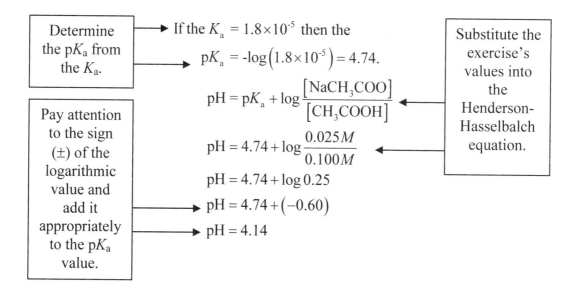

| Determine the pK_a from the K_a. | If the $K_a = 1.8 \times 10^{-5}$ then the | Substitute the exercise's values into the Henderson-Hasselbalch equation. |

$$pK_a = -\log(1.8 \times 10^{-5}) = 4.74.$$

$$pH = pK_a + \log\frac{[NaCH_3COO]}{[CH_3COOH]}$$

Pay attention to the sign ($\pm$) of the logarithmic value and add it appropriately to the pK_a value.

$$pH = 4.74 + \log\frac{0.025\,M}{0.100\,M}$$

$$pH = 4.74 + \log 0.25$$

$$pH = 4.74 + (-0.60)$$

$$pH = 4.14$$

Module 8
Functional Groups and Organic Nomenclature

Introduction

This module describes the various functional groups found in organic chemistry and addresses the IUPAC system for naming organic compounds. Atoms or groups of atoms that undergo predictable chemical reactions are called *functional groups*.

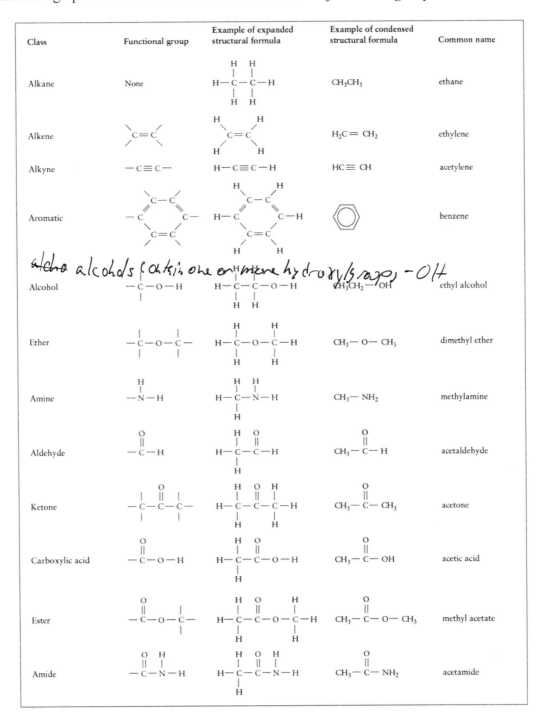

Class	Functional group	Example of expanded structural formula	Example of condensed structural formula	Common name
Alkane	None	H—C—C—H (with H's)	CH_3CH_3	ethane
Alkene	C=C	H₂C=CH₂ structure	$H_2C = CH_2$	ethylene
Alkyne	—C≡C—	H—C≡C—H	$HC \equiv CH$	acetylene
Aromatic	ring	ring	benzene ring	benzene
Alcohol	—C—O—H	H—C—C—O—H	$CH_3CH_2—OH$	ethyl alcohol
Ether	—C—O—C—	H—C—O—C—H	$CH_3—O—CH_3$	dimethyl ether
Amine	—N—H	H—C—N—H	$CH_3—NH_2$	methylamine
Aldehyde	—C(=O)—H	H—C—C(=O)—H	$CH_3—C(=O)—H$	acetaldehyde
Ketone	—C—C(=O)—C—	H—C—C(=O)—C—H	$CH_3—C(=O)—CH_3$	acetone
Carboxylic acid	—C(=O)—O—H	H—C—C(=O)—O—H	$CH_3—C(=O)—OH$	acetic acid
Ester	—C(=O)—O—C—	H—C—C(=O)—O—C—H	$CH_3—C(=O)—O—CH_3$	methyl acetate
Amide	—C(=O)—N—H	H—C—C(=O)—N—H	$CH_3—C(=O)—NH_2$	acetamide

(handwritten annotation across Alcohol row:) alcohols (contains one or more hydroxyl group) −OH

69

General Guide to the IUPAC System of Organic Nomenclature

There are four basic steps to naming any organic compound:

1. Identify the parent chain

The parent chain is the longest continuous chain of carbon atoms. The name of the parent chain indicates the number of carbons in the chain. If there are functional groups present in the molecule, the parent chain must include the priority functional group and the name of the parent chain is modified to indicate the functional group present.

Number of carbon atoms	Name	Molecular formula	Structure of normal isomer
1	methane	CH_4	CH_4
2	ethane	C_2H_6	CH_3CH_3
3	propane	C_3H_8	$CH_3CH_2CH_3$
4	butane	C_4H_{10}	$CH_3CH_2CH_2CH_3$
5	pentane	C_5H_{12}	$CH_3CH_2CH_2CH_2CH_3$
6	hexane	C_6H_{14}	$CH_3CH_2CH_2CH_2CH_2CH_3$
7	heptane	C_7H_{16}	$CH_3CH_2CH_2CH_2CH_2CH_2CH_3$
8	octane	C_8H_{18}	$CH_3CH_2CH_2CH_2CH_2CH_2CH_2CH_3$
9	nonane	C_9H_{20}	$CH_3CH_2CH_2CH_2CH_2CH_2CH_2CH_2CH_3$
10	decane	$C_{10}H_{22}$	$CH_3CH_2CH_2CH_2CH_2CH_2CH_2CH_2CH_2CH_3$

2. Number the carbons of the parent chain

Begin numbering alkane chains in the direction that will give the first branching carbon the lowest number. If numbering the chain in either direction leads to the same lowest number for the first substituent on the chain, number in the direction that gives the lowest possible number to the next second substituent on the chain, and so on. If the chain contains a functional group, number the chain in the direction that gives the lowest number to the priority functional group. Functional groups are shown in order of decreasing priority in the following table.

Functional group	Name as suffix	Name as prefix
Principal groups		
Carboxylic acids	-oic acid -carboxylic acid	carboxy
Acid anhydrides	-oic anhydride -carboxylic anhydride	—
Esters	-oate -carboxylate	alkoxycarbonyl
Thioesters	-thioate -carbothioate	alkylthiocarbonyl
Acid halides	-oyl halide -carbonyl halide	halocarbonyl
Amides	-amide -carboxamide	carbamoyl
Nitriles	-nitrile -carbonitrile	cyano
Aldehydes	-al -carbaldehyde	oxo
Ketones	-one	oxo
Alcohols	-ol	hydroxy
Phenols	-ol	hydroxy
Thiols	-thiol	mercapto
Amines	-amine	amino
Imines	-imine	imino
Ethers	ether	alkoxy
Sulfides	sulfide	alkylthio
Disulfides	disulfide	—
Alkenes	-ene	—
Alkynes	-yne	—
Alkanes	-ane	—
Subordinate groups		
Azides	—	azido
Halides	—	halo
Nitro compounds	—	nitro

[a]Principal groups are listed in order of decreasing priority; subordinate groups have no priority order.

3. **Identify and locate branches and/or substituents**

   Carbon atoms attached to the parent carbon chain are called *branches* or *side-chains*. Other atoms or groups attached to the parent carbon chain are commonly called *substituents,* although substituents can be applied more generally to both carbon branches and other atoms or groups. Hydrocarbon branches are named by replacing the *ane* ending with *yl.* They are designated by writing the number of the parent-chain carbon where the branch occurs followed by a hyphen and the branch name. If two or more substituents are the same, the prefixes "di", "tri", "tetra", and so forth, are used to indicate how many identical substituents are present.

Parent alkane	Structure of parent alkane	Structure of alkyl group	Name of alkyl group
methane	CH_4	CH_3-	methyl
ethane	CH_3CH_3	CH_3CH_2-	ethyl
propane	$CH_3CH_2CH_3$	$CH_3CH_2CH_2-$	propyl
		CH_3CHCH_3	isopropyl
n-butane	$CH_3CH_2CH_2CH_3$	$CH_3CH_2CH_2CH_2-$	butyl
		$CH_3CH_2CHCH_3$	*sec*-butyl (secondary-butyl)[a]
isobutane	CH_3CHCH_3 (with CH_3 branch)	CH_3CHCH_2- (with CH_3 branch)	isobutyl
		CH_3CCH_3 (with CH_3 above and below)	*t*-butyl (tertiary-butyl)[a]

4. **Assemble the name**

   Branches and substituents are written in alphabetical order preceding the name of the parent carbon chain. The prefixes "di", "tri", "tetra", "sec", and "*tert*" are ignored in alphabetizing substituents. Numbers are separated from names by hyphens and from numbers by commas.

Properties and Nomenclature of Alkanes (C_nH_{2n+2})

Alkanes are the simplest organic compounds because they consist only of hydrogens and carbons joined by single bonds. They are nonpolar compounds whose intermolecular attractions are dominated by London dispersion forces. Their boiling points increase with increased molecular weight. For a given molecular formula their boiling points tend to decrease with increased branching. Alkanes are insoluble in polar solvents, like water, but soluble in other nonpolar solvents.

Sample Exercises

Naming Alkanes

1. Use the IUPAC system to name the following alkane.

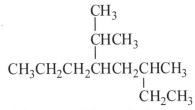

The correct answer is: 5-isopropyl-3-methyloctane

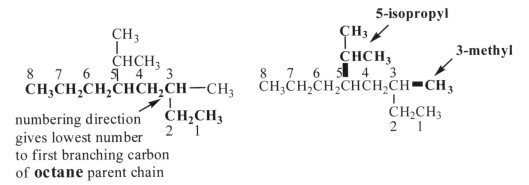

2. Use the IUPAC system to name the following alkane.

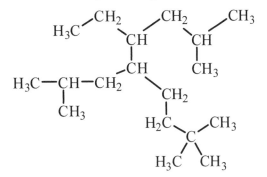

The correct answer is: 6-ethyl-5-isobutyl-2,2,8-trimethylnonane

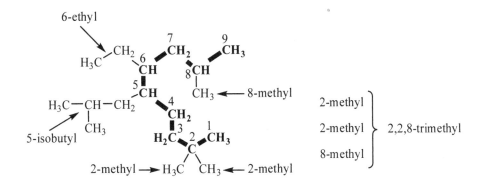

There are three methyl branches on the parent nonane chain. Three identical branches are designated by the prefix "tri". Every substituent must be located on the parent chain. Therefore, the parent nonane chain contains a 2-methyl branch, another 2-methyl branch, and an 8-methyl branch. The names and locations of these three identical branches are consolidated in the IUPAC name as "2,2,8-trimethyl." Also, notice that any one of the three methyl groups on C2 could be included in the parent chain with the other two being designated as branches. The name would be unchanged. The same is true for the two methyl groups on C8.

It is incorrect to number the parent nonane chain from the other end, as shown below. Be careful! This is a common mistake when more than one branch is located on the same carbon of the parent chain. If numbering the parent chain from either end gives the same number for the first substituent, the chain must be numbered in the direction that will give the next substituent the lowest possible number, and so forth. When this chain is numbered correctly, the next substituent is located at C2, rather than at C4 when the chain is numbered incorrectly.

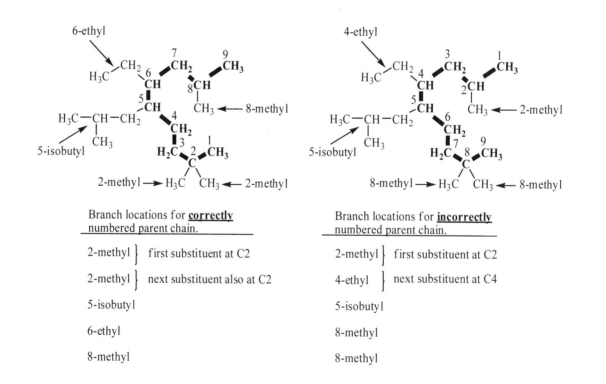

Branch locations for **correctly** numbered parent chain.

2-methyl	}	first substituent at C2
2-methyl	}	next substituent also at C2
5-isobutyl		
6-ethyl		
8-methyl		

Branch locations for **incorrectly** numbered parent chain.

2-methyl	}	first substituent at C2
4-ethyl	}	next substituent at C4
5-isobutyl		
8-methyl		
8-methyl		

3. Use the IUPAC system to name the following alkane.

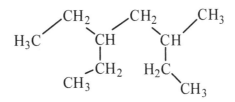

The correct answer is: 3-ethyl-5-methylheptane

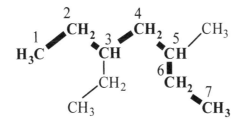

When numbered from one end, the heptane chain contains 3-ethyl and 5-methyl branches. When numbered from the other end, the heptane chain contains 5-ethyl and 3-methyl branches. The same substituent numbers are obtained in both directions. Consequently, the chain must be numbered in the direction that gives the ethyl branch, which is listed first alphabetically, the lower number.

Properties and Nomenclature of Cycloalkanes (C_nH_{2n})

Cycloalkanes are saturated hydrocarbons, like alkanes, with their carbon atoms bonded to form a ring. Their properties are virtually the same as those for chain alkanes.

Cycloalkanes with one substituent are named as un-numbered cycloalkanes with alkyl substituents, unless the side-chain has more carbons than the ring. In that case, the chain is the parent hydrocarbon with a cycloalkyl substituent. If the ring has two alkyl substituents, they are cited in alphabetical order and the ring is numbered to give the number-1 position to the first substituent cited. For rings with three or more alkyl substituents, the substituents are cited in alphabetical order and the ring is numbered either clockwise or counterclockwise to give the lowest sequence of numbers for the substituents. If the ring has an attached high priority functional group, the ring-carbon bearing the priority functional group receives number-1 and the ring is numbered either clockwise or counterclockwise to give the next highest priority functional group the lowest number.

Sample Exercises

Naming Cycloalkanes
4. *Use the IUPAC system to name the following cycloalkane.*

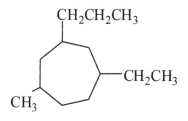

The correct answer is: 1-ethyl-5-methyl-3-propylcycloheptane

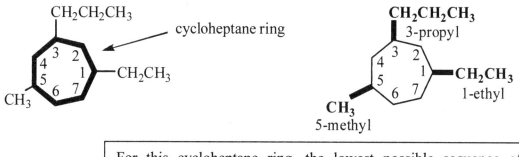

For this cycloheptane ring, the lowest possible sequence of numbers for the branches is 1,3,5. The ring is numbered counterclockwise starting from the ethyl branch because ethyl is cited first alphabetically.

5. *Use the IUPAC system to name the following cycloalkane.*

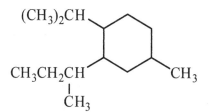

The correct answer is: 2-*sec*-butyl-1-isopropyl-4-methylcyclohexane

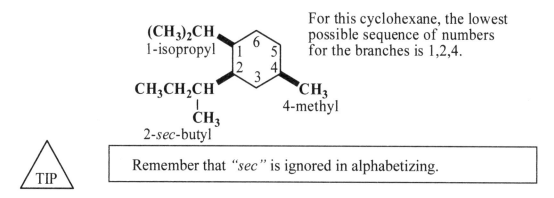

For this cyclohexane, the lowest possible sequence of numbers for the branches is 1,2,4.

Remember that *"sec"* is ignored in alphabetizing.

6. *Use the IUPAC system to name the following cycloalkane.*

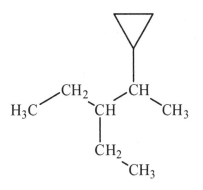

The correct answer is: 2-cyclopropyl-3-ethylpentane

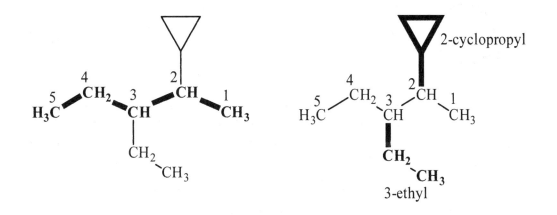

If a compound contains two possibilities for the longest chain, the parent chain is the one that contains the greatest number of branches.

If the chain contains more carbon atoms than the ring, the compound is named as an alkane chain with a cycloalkyl substituent.

77

Cis/Trans Nomenclature for Cycloalkanes

Cycloalkanes are alkanes with their carbon atoms in a ring. The ring confers "sidedness" to the molecule. Substituents located on the same side of the ring are *cis* to each other. Substituents located on opposite sides of the ring are *trans* to each other.

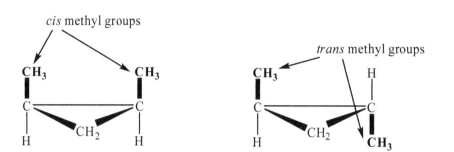

To name a disubstituted cycloalkane, compare the two substituents on the ring. Write *cis* preceding the name if the substituents are *cis* and write *trans* preceding the name if the substituents are *trans*.

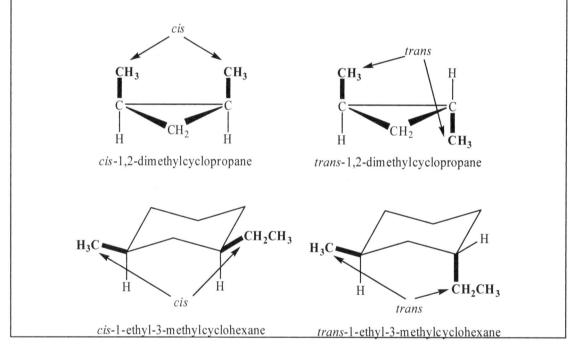

cis-1,2-dimethylcyclopropane   *trans*-1,2-dimethylcyclopropane

cis-1-ethyl-3-methylcyclohexane   *trans*-1-ethyl-3-methylcyclohexane

7. *Use the IUPAC system to name the following cycloalkane.*

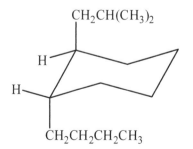

The correct answer is: *trans*-1-butyl-2-isobutylcyclohexane

Hydrogen is directed
down from ring carbon

Hydrogen is directed
up from ring carbon

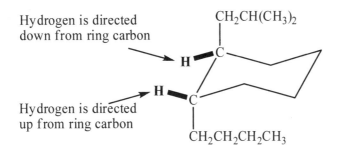

To determine if hydrogens are *cis* or *trans* to each other on a cyclohexane chair, look at the ring carbon to which each hydrogen is attached. In addition to the two ring bonds, each carbon has two remaining bonds, one directed up from the ring carbon and one directed down from the ring carbon. If the hydrogens are both occupying bonds directed up or down from the ring carbons, they are *cis*. Alternatively, if one hydrogen is directed up and the other down, they are *trans*.

Properties and Nomenclature of Alkenes (C_nH_{2n})

Alkenes are hydrocarbons that include one or more double bonds. The physical properties of alkenes are similar to those of alkanes. Both groups of compounds are nonpolar whose intermolecular attractions are dominated by London dispersion forces. The reactions of alkenes predominantly involve the π-bond.

Alkenes, like cycloalkanes, exhibit "sidedness." The orientation of the carbon atoms of the parent chain determines whether an alkene is *cis* or *trans*. If the parent chain continues on the same side of the double bond it is a *cis* alkene; alternatively, if the parent chain continues on the opposite side of the double bond, it is a *trans* alkene.

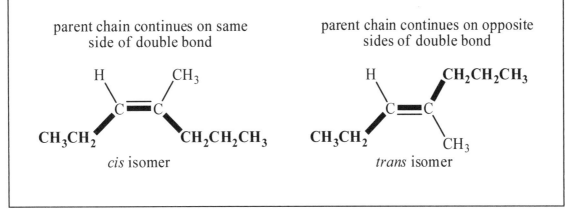

To name an alkene using the IUPAC system:

1. Identify the parent chain. The parent chain is the longest continuous carbon chain <u>that contains both carbons of the double bond</u>. The suffix "ane" is changed to "ene" to indicate the presence of the double bond in the parent chain. Number the parent chain in the direction that gives the carbons of the double bond the lowest possible numbers and locate the double bond in the parent chain by writing the lower of the two numbers. For example, 1-hexene signifies that the double bond is between C1 and C2 of a hexene chain.

1. If the parent chain contains two double bonds, the suffix is "diene". For example, 1,3-pentadiene signifies that the five carbon parent chain contains a double bond between C1 and C2 and another double bond between C3 and C4.

2. Substituents are cited in alphabetical order and the appropriate number is assigned to each substituent.

3. To indicate *cis/trans* isomers, write *cis* or *trans* before citing and locating substituents.

Sample Exercise

Naming Alkenes

8. Use the IUPAC system to name the following alkene.

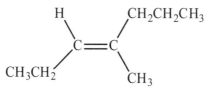

The correct answer is: *trans*-**4-methyl-3-heptene**

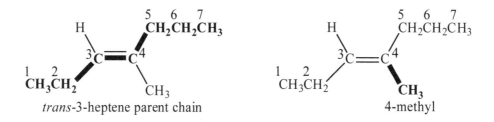

trans-3-heptene parent chain   4-methyl

9. Use the IUPAC system to name the following alkene.

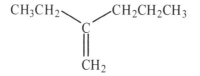

The correct answer is: 2-ethyl-1-pentene

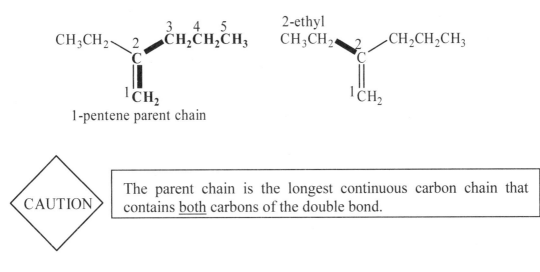

1-pentene parent chain   2-ethyl

The parent chain is the longest continuous carbon chain that contains both carbons of the double bond.

Properties and Nomenclature of Cycloalkenes (C_nH_{2n-2})

Cycloalkenes and alkenes exhibit almost identical physical properties. They are nonpolar compounds whose intermolecular forces are dominated by London dispersion forces.

A number is not needed to locate the position of the double bond in a cyclic alkene. The ring is always numbered so that the double bond is between C1 and C2. To locate a substituent, number the carbons of the ring in the direction (clockwise or counterclockwise) that gives the lowest number(s) for the substituent(s).

Sample Exercise

Naming Cycloalkenes
10. Use the IUPAC system to name the following cycloalkene.

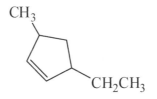

The correct answer is: 3-ethyl-5-methylcyclopentene

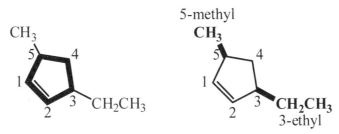

cyclopentene ring with double bond
located between C1 and C2

81

TIPS

If this cyclopentene ring is numbered in either the clockwise or counterclockwise direction the substituents will be located at carbons 3 and 5. The ring is numbered in the counterclockwise direction to give the ethyl substituent (the first cited substituent alphabetically) the lowest number.

Properties and Nomenclature of Alkynes (C_nH_{2n-2})

Alkynes are hydrocarbons that include at least one triple bond. Like alkanes and alkenes, alkynes are nonpolar compounds whose intermolecular attractions are dominated by London Dispersion forces. As with alkenes, the π-bond is the site of almost all alkyne reactions.

Sample Exercise

Naming Alkynes
11. Use the IUPAC system to name the following alkyne.

$$CH_3CH_2-C\equiv C-\overset{\overset{\displaystyle CH_3}{|}}{\underset{\underset{\displaystyle CH_3}{|}}{C}}-CH_3$$

The correct answer is: 2,2-dimethyl-3-hexyne

Properties and Nomenclature of Benzene and its Derivatives

Benzene (C_6H_6) belongs to a group of hydrocarbons called aromatic hydrocarbons, historically named because many benzene derivatives possess distinctive aromas. Benzene is nonpolar, insoluble in water, and has a low melting point. It is represented by writing two resonance contributing structures which differ only in the position of the double bonds. A hybrid of these two contributing structures is a good description for the real structure of benzene.

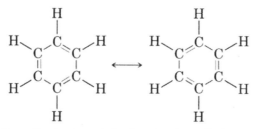

Alternative Lewis contributing structures for benzene

Monosubstituted alkylbenzenes are usually named as benzene derivatives. The IUPAC system also retains common names for other monosubstituted benzenes.

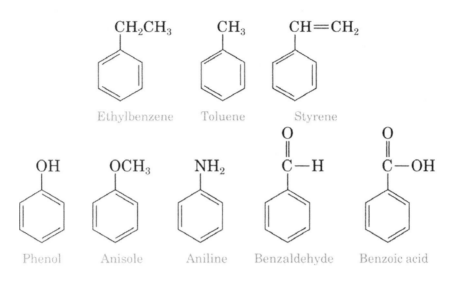

Ethylbenzene   Toluene   Styrene

Phenol   Anisole   Aniline   Benzaldehyde   Benzoic acid

There are three isomers possible for disubstituted benzenes.

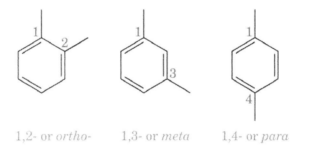

1,2- or *ortho*-   1,3- or *meta*   1,4- or *para*

When one of the substituents imparts a special name (such as -CH₃, -NH₂, -COOH), the molecule is named as a derivative of that compound and the special substituent occupies ring position number 1.

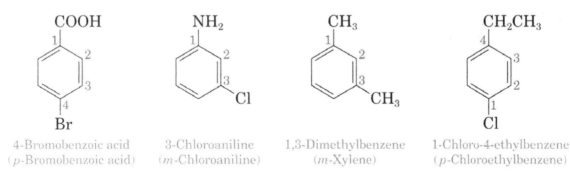

4-Bromobenzoic acid   3-Chloroaniline   1,3-Dimethylbenzene   1-Chloro-4-ethylbenzene
(*p*-Bromobenzoic acid)   (*m*-Chloroaniline)   (*m*-Xylene)   (*p*-Chloroethylbenzene)

If there are three or more substituents on the benzene ring, the locations of the substituents are specified by numbers and the ring is numbered so that the smallest set of numbers is obtained.

Sample Exercises

Naming Benzene Derivatives
12. Use the IUPAC system to name the following substituted benzenes.

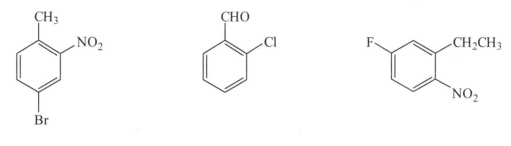

The correct answers are:

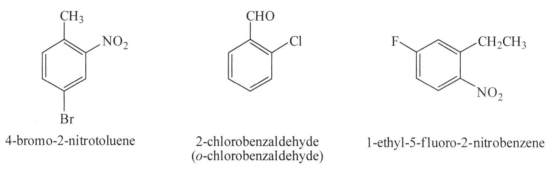

4-bromo-2-nitrotoluene   2-chlorobenzaldehyde   1-ethyl-5-fluoro-2-nitrobenzene
                             (*o*-chlorobenzaldehyde)

Properties and Nomenclature of Alcohols (*R-OH*)

Alcohols are compounds that have one or more hydroxyl groups (-OH) in the compound. The inclusion of the hydroxyl group renders low molecular weight alcohols soluble in water. High molecular weight alcohols become increasingly water-soluble with the inclusion of additional hydroxyl groups. Alcohols boil at higher temperatures due to hydrogen bonding.

The –*e* suffix is replaced with –*ol* to show that the compound is an alcohol.

Sample Exercise

Naming Alcohols
13. Use the IUPAC system to name the following alcohol

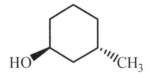

The correct answer is: *trans*-3-methylcyclohexanol

Properties and Nomenclature of Ethers (*R-O-R'*)

Ethers are organic compounds containing two carbon chains or rings joined by an oxygen atom. Due to their polar covalent carbon-oxygen bonds, ethers are slightly more soluble in water than hydrocarbons. Their boiling points are similar to hydrocarbons with comparable molecular weights.

If the alkyl groups of the ether have common names, the ether is named by citing the alkyl groups alphabetically followed by "ether", as in *ethyl methyl ether* or *methyl propyl ether*. If one or both of the alkyl groups cannot be named commonly, the larger chain or ring is designated the parent and the other group is named as an alkoxy group, such as *methoxy* or *isopropoxy*.

Sample Exercise

Naming Ethers
14. Use the IUPAC system to name the following ether.

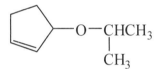

The correct answer is: 3-isopropoxycyclopentene

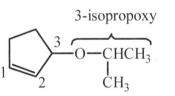

3-isopropoxy

The designation 3-isopropoxy locates the isopropoxy substituent at C3 of the ring and also locates the substituent relative to the double bond.

Properties and Nomenclature of Thiols (*R-SH*)

Thiols are organic compounds that contain a sulfhydryl group, -SH. The difference in electronegativities between sulfur and hydrogen is so small that the S-H bond is basically non-polar. Therefore, thiols have boiling points similar to hydrocarbons of the same molecular weight and are insoluble in water. The most notable characteristic of thiols is their stench which is why thiols are added to natural gas to help detect gas leaks.

Thiols are named by selecting the longest carbon chain or largest ring that contains the sulfhydryl group and adding the suffix *thiol* to the name of the parent alkane. Thus, $CH_3CH_2CH_2SH$ is named *1- propanethiol*.

Naming Thiols
15. Use the IUPAC system to name the following thiol.

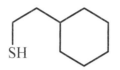

The correct answer is 2-cyclohexylethanethiol.

	Although the cyclohexane ring contains more carbons than its attached chain, the attached chain contains the sulfhydryl group. The chain is numbered to give the carbon with the sulfydryl group the lowest number. It is not necessary to write the number 1 to designate the position of the sulfhydryl group. 2-cyclohexyl unambiguously describes the relative positions of the cyclohexyl substituent and the sulfhydryl group.

Properties and Nomenclature of Amines ($R\text{-}NH_2$, $R\text{-}NHR'$, $R\text{-}NR'R''$)

Amines are organic compounds that include nitrogen atoms. All amines form hydrogen bonds with water and are more soluble in water than hydrocarbons of comparable molecular weight. Their most significant chemical property is their basicity. Amines are classified as primary, secondary, or tertiary depending on the number of carbon groups attached to the nitrogen. Primary (1°) amines have one carbon group and two hydrogens bonded to the nitrogen. They form hydrogen bonds with each other, although they are weaker than hydrogen bonds found in alcohols. Secondary (2°) amines have two carbon groups and one hydrogen atom bonded to the nitrogen and also form hydrogen bonds with one another. Tertiary (3°) amines have three carbon groups attached to the nitrogen and no hydrogens. Tertiary amines cannot form hydrogen bonds with one another.

Amines are named by selecting the longest continuous chain or largest ring that contains the amine group and writing the suffix *amine* after the parent name.

Sample Exercise

Naming Amines
16. Use the IUPAC system to name the following amine.

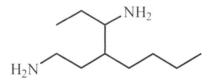

The correct answer is: 3-butyl-1,4-hexanediamine

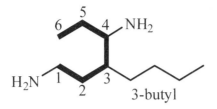

parent chain must contain both amine groups

3-butyl

Properties and Nomenclature of Aldehydes (*RCHO, HCHO*) and Ketones (*RCOR′*)

Aldehydes and ketones are classes of organic compounds that include a carbon-oxygen double bond. This structure is called a *carbonyl group*. Ketones contain two carbon

carbonyl group

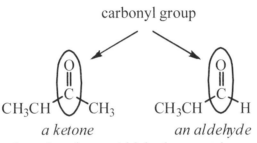

a ketone   *an aldehyde*

groups attached to the carbonyl carbon. Aldehydes contain one carbon group and one hydrogen atom attached to the carbonyl carbon, or two hydrogen atoms attached to the carbonyl carbon.

The carbonyl group is a determining factor in the physical and chemical properties of aldehydes and ketones. The carbonyl group is a dipole due to the difference in electronegativities between the carbon and oxygen atoms. The carbon atom carries a partial positive charge and the oxygen atom carries a partial negative charge. The carbonyl carbon of one molecule can associate with the carbonyl oxygen of another molecule. This dipole-dipole interaction is much stronger than the London Forces found in alkanes but weaker than the hydrogen bonding found in alcohols.

Aldehydes are named by appending the suffix -*al* to the parent chain or ring. Ketones are named by appending the suffix -*one* to the parent chain or ring.

Sample Exercise

Naming Aldehydes and Ketones
17. Use the IUPAC system to name the following ketone.

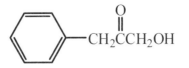

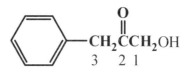

$$\overset{3\quad2\quad1}{—CH_2\overset{\overset{\displaystyle O}{\|}}{C}CH_2OH}$$

The correct answer is: 1-hydroxy-3-phenyl-2-propanone

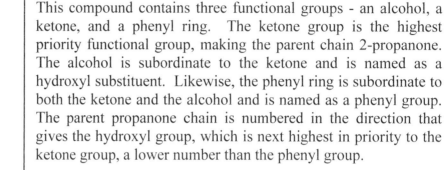

This compound contains three functional groups - an alcohol, a ketone, and a phenyl ring. The ketone group is the highest priority functional group, making the parent chain 2-propanone. The alcohol is subordinate to the ketone and is named as a hydroxyl substituent. Likewise, the phenyl ring is subordinate to both the ketone and the alcohol and is named as a phenyl group. The parent propanone chain is numbered in the direction that gives the hydroxyl group, which is next highest in priority to the ketone group, a lower number than the phenyl group.

Properties and Nomenclature of Carboxylic Acids (_RCOOH_)

Carboxylic acids are organic compounds that contain a carbonyl with a hydroxyl group attached to it. This composite functional group is called a **carboxyl group** and can be represented in several ways.

The carboxyl group has three very polar bonds (C=O, C-O, and O-H) and it is this polarity that determines the physical and chemical properties of the compound. Carboxylic acids have much stronger intermolecular attractions than alcohols and are more soluble in water.

Carboxylic acids are named by appending the suffix _-oic acid_ to the parent chain or _carboxylic acid_ to the parent ring.

Sample Exercise

Naming Carboxylic Acids
18. Use the IUPAC system to name the following carboxylic acid.

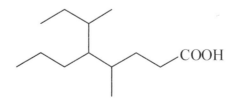

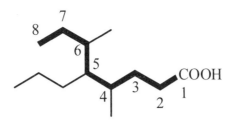

The correct answer is: 4,6-dimethyl-5-propyloctanoic acid.

The parent chain must include the carbon of the carboxyl group. There are two eight-carbon chains in this carboxylic acid. When the compound contains two chains with the same number of carbon atoms, the parent chain is the one that contains the most branches.

Properties and Nomenclature of Acid Anhydrides [(RCO)$_2$O]

Acid anhydrides are carboxylic acid derivatives where two carbonyl groups are attached to the same oxygen. Acid anhydrides can be formed by *dehydrating* two carboxylic acids. If the two acids are the same, then the anhydride is symmetrical. If the two acids are different, then the anhydride is asymmetrical. Acid anhydrides react readily with water to yield carboxylic acids.

Acid anhydrides are named as the derivative of the acid or acids used to prepare them. The word *acid* is dropped from the name and replaced with *anhydride*. Symmetrical anhydrides omit the *di* prefix. Below are some examples of acid anhydrides.

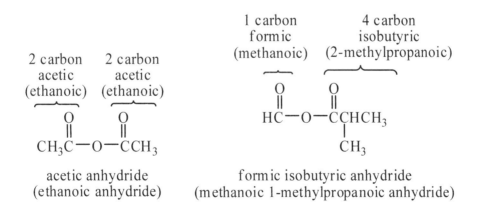

Properties and Nomenclature of Esters (*RCO₂R′*)

Esters are carboxylic acid derivatives that have an alkoxyl group (-OR) attached to the carbonyl carbon in the place of the hydroxyl (-OH) group.

Esters are named as derivatives of a parent carboxylic acid. The name of the alkoxyl group is placed before the name of the parent acid. The *–ic acid* is dropped from the parent name of the acid and the suffix *–ate* is used. Below are some examples of esters.

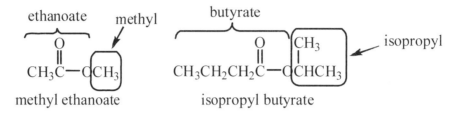

Properties and Nomenclature of Amides (*RCONH₂, ROCNHR′, RCONR′₂*)

Amides are organic compounds that include a carbonyl group connected to a nitrogen atom. The nitrogen atom then may be bonded to hydrogens or other carbon groups. Amides are named by removing the *–oic acid* suffix from the analogous carboxylic acid and appending *–amide*. If carbon groups are connected to the nitrogen of the amide, the groups are named and its location on the nitrogen is indicated by *N-*. Below are some examples of amides.

$$
\underset{\text{ethanamide}}{CH_3\overset{\overset{\displaystyle O}{\|}}{C}-NH_2}
\qquad\qquad
\underset{N,N\text{-dimethylpropanamide}}{CH_3CH_2\overset{\overset{\displaystyle O}{\|}}{C}-\underset{\underset{\displaystyle CH_3}{|}}{N}-CH_3}
$$

Sample Exercises

Naming Carboxylic Acid Derivatives
19. Use the IUPAC system to name the following carboxylic acid derivatives

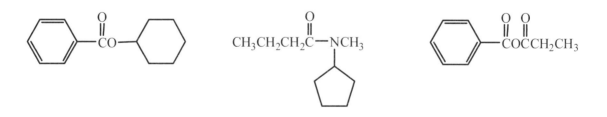

The correct answers are:

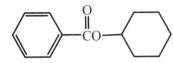

cyclohexyl benzoate

$CH_3CH_2CH_2\overset{\displaystyle O}{C}-NCH_3$

N-cyclopentyl-N-methylbutanamide

benzoic propanoic anhydride

$\overset{\displaystyle O\ \ O}{COCCH_2CH_3}$

Module 9
Stereoisomerism

Introduction

This module describes stereoisomers and their characteristics. The goals of this module are to show you how to:
1. visualize three dimensional objects given two dimensional drawings
2. recognize stereocenters
3. understand the relationships between enantiomers, diastereomers, and meso compounds
4. discern the configuration at a chiral center
5. understand the interaction of plane-polarized light with chiral molecules

CAUTION

| You will need access to a molecular modeling kit. It is **highly** recommended that you build the molecules in the examples to see what is being described. |

Module 9 Key Concepts

Definitions
- *a.* *chiral* – objects that are not superimposable on their mirror images.
- *b.* *achiral* – objects that are superimposable on their mirror images.
- *c.* *stereocenter* – a tetrahedral atom with 4 different atoms or groups. A carbon atom with four different attached substituents is a stereocenter. A nitrogen atom with three different substituents and a lone pair of electrons is a stereocenter.
- *d.* *enantiomers* – a pair of stereoisomers that are nonsuperimposable mirror images.
- *e.* *diastereomers* – stereoisomers that are not mirror images of each other.
- *f.* *meso compounds* – achiral compounds that contain stereocenters.
- *g.* *racemic mixture* – a 50:50 mixture of enantiomers.
- *h.* *Optical activity* – property of a compound to rotate the plane of polarized light. Achiral compounds and racemic mixtures do not rotate the plane of polarized light.
- *i.* *R/S Nomenclature of stereocenters*
  IUPAC system of nomenclature that utilizes the *Cahn-Ingold-Prelog* rules to prioritize the substituents on a stereocenter and assign configuration.

Visualizing Molecules

One hurdle textbook authors and chemistry instructors face is representing 3-dimensional molecules on a 2-dimensional page. It is important that you understand the different ways molecules are represented so that you understand what an illustration is attempting to convey.

One method of representing molecules is with dashes and wedges as shown below.

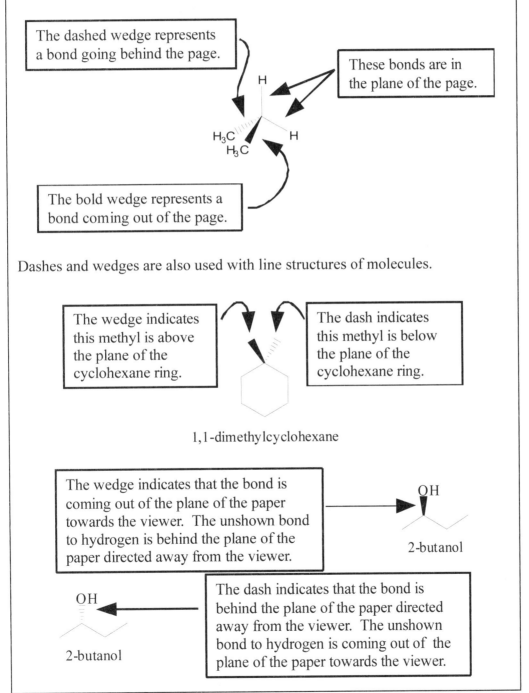

The dashed wedge represents a bond going behind the page.

These bonds are in the plane of the page.

The bold wedge represents a bond coming out of the page.

Dashes and wedges are also used with line structures of molecules.

The wedge indicates this methyl is above the plane of the cyclohexane ring.

The dash indicates this methyl is below the plane of the cyclohexane ring.

1,1-dimethylcyclohexane

The wedge indicates that the bond is coming out of the plane of the paper towards the viewer. The unshown bond to hydrogen is behind the plane of the paper directed away from the viewer.

2-butanol

2-butanol

The dash indicates that the bond is behind the plane of the paper directed away from the viewer. The unshown bond to hydrogen is coming out of the plane of the paper towards the viewer.

Fischer projections are commonly used to represent carbohydrates. In Fischer projections the vertical bonds are going into the plane of the paper away from the viewer and the horizontal bonds are coming out of the plane of the paper towards the viewer.

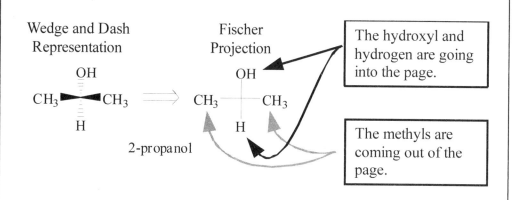

The hydroxyl and hydrogen are going into the page.

The methyls are coming out of the page.

Enantiomers

Stereochemistry involves the orientation of atoms or groups in three-dimensional space. Some molecules differ from one another only in their three-dimensional orientation, just as a right and left hand have the same features but are not identical. In fact, right and left hands are nonsuperimposable mirror images of each other. Some molecules are also *handed*, or in other words, they are nonsuperimposable mirror images of each other.

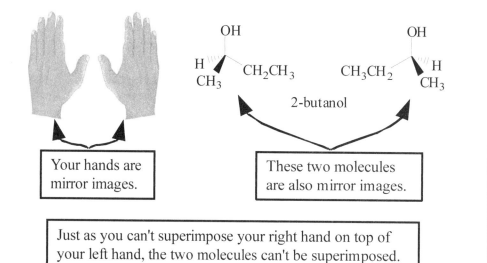

Your hands are mirror images.

These two molecules are also mirror images.

Just as you can't superimpose your right hand on top of your left hand, the two molecules can't be superimposed.

Take a moment to build two molecular models of 2-butanol. If you happen to end up with two molecules that can be superimposed, simply swap the positions of the hydrogen and hydroxyl on one model. Verify that the two models are mirror images and that you can not superimpose one over the other. No matter how you try, two of the substituents will be in the wrong place.

In chemical nomenclature, when two molecules that are mirror images of each other cannot be superimposed onto one another, the objects are said to be **chiral**, from Greek *cheir, "hand"*. If an object and its mirror image can be superimposed then the molecule is said to be **achiral**.

2-Butanol is an example of a chiral molecule. It exists in two stereoisomeric forms. Nonsuperimposable stereoisomers of a molecule are called **enantiomers**. Similar to *cis/trans* isomers, enantiomers have the same molecular formulas and connectivities of atoms but different arrangements in space.

Stereocenters

2-Butanol is chiral because of the second carbon in the chain. C2 is a tetrahedral carbon with four different substituents; a hydrogen, a hydroxyl group, a methyl group, and an ethyl group. A tetrahedral carbon that is bonded to four different atoms or groups is called a **stereocenter**. More generally, any tetrahedral center that has four different substituents around it is a stereocenter.

This carbon is a stereocenter because it is bonded to four different atoms or groups: a chlorine atom, a methyl group, an ethyl group, and a hydrogen atom.

C4 is a stereocenter with four different attached atoms or groups: a hydrogen atom, a chlorine atom, the ring clockwise from C4, and the ring counterclockwise from C4. Proceeding clockwise around the ring from C4 is a CH$_2$ at C5 and a CH$_2$ at C6. Proceeding counterclockwise around the ring from C4 is a CH$_2$ at C3 and a double-bonded carbon at C2.

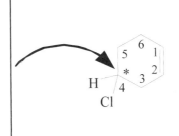

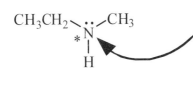

The nitrogen is a stereocenter because it is surrounded by four different atoms or groups: a hydrogen atom, a methyl group, an ethyl group, and a lone pair of electrons.

All molecules containing a single stereocenter exist as a chiral pair of enantiomers.

Diastereomers

If there are two or more stereocenters in a molecule, then more stereoisomers are possible. Consider the four stereoisomers of 2-bromo-3-chlorobutane.

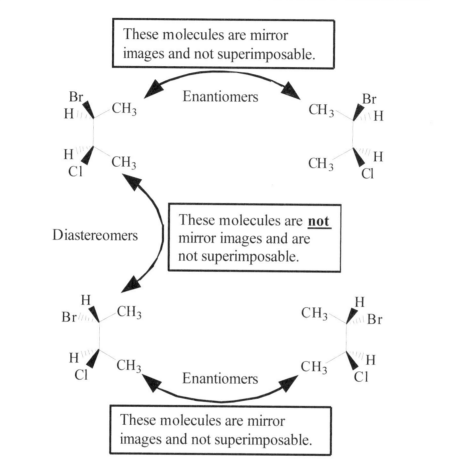

2-bromo-3-chlorobutane exists as two enantiomeric pairs, meaning that there are two nonsuperimposable pairs of stereoisomers. Although all four stereoisomers of 2-bromo-3-chlorobutane have the same molecular formulas and the same connectivities of atoms, the two enantiomeric pairs differ from each other because they are mirror images of each other. Nonsuperimposable stereoisomers that are not mirror images of each other are called *diastereomers*.

Meso Compounds

Some compounds containing stereocenters are achiral. Consider the stereoisomers of 2,3-dibromobutane.

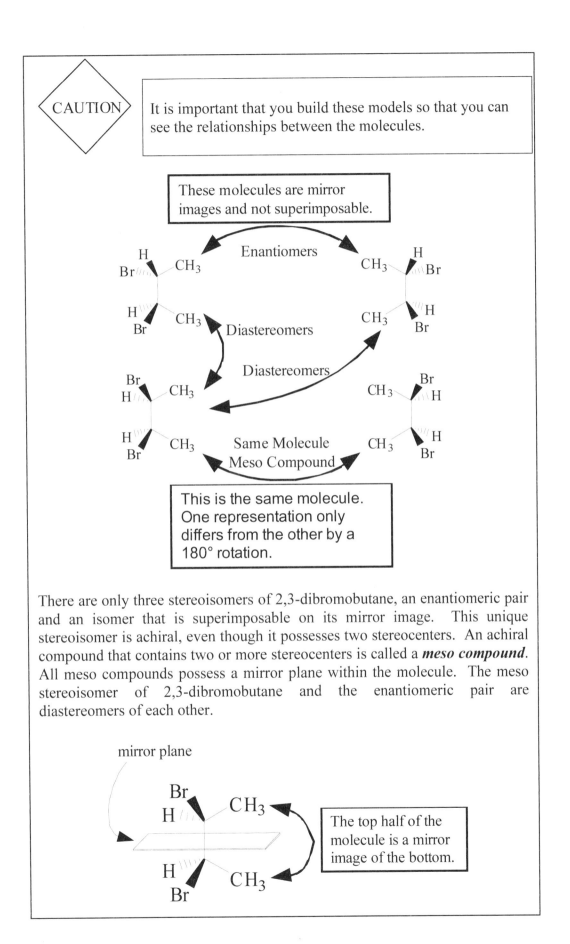

CAUTION It is important that you build these models so that you can see the relationships between the molecules.

These molecules are mirror images and not superimposable.

Enantiomers

Diastereomers

Diastereomers

Same Molecule
Meso Compound

This is the same molecule. One representation only differs from the other by a 180° rotation.

There are only three stereoisomers of 2,3-dibromobutane, an enantiomeric pair and an isomer that is superimposable on its mirror image. This unique stereoisomer is achiral, even though it possesses two stereocenters. An achiral compound that contains two or more stereocenters is called a *meso compound*. All meso compounds possess a mirror plane within the molecule. The meso stereoisomer of 2,3-dibromobutane and the enantiomeric pair are diastereomers of each other.

mirror plane

The top half of the molecule is a mirror image of the bottom.

Specifying Configurations

The IUPAC system of organic nomenclature defines a method for uniquely identifying stereoisomers. First, the substituents on the stereocenter are prioritized based upon the *Cahn-Ingold-Prelog* rules.

1. Prioritize the atoms directly attached to the stereocenter based on their atomic number. The largest atomic number gets the highest priority.

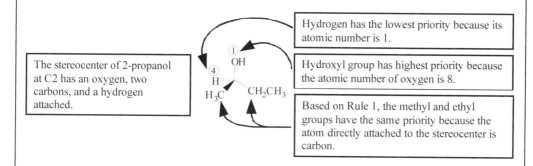

The stereocenter of 2-propanol at C2 has an oxygen, two carbons, and a hydrogen attached.

Hydrogen has the lowest priority because its atomic number is 1.

Hydroxyl group has highest priority because the atomic number of oxygen is 8.

Based on Rule 1, the methyl and ethyl groups have the same priority because the atom directly attached to the stereocenter is carbon.

2. If the atoms attached to the stereocenter are the same element, then compare the atoms one bond further removed from the stereocenter until a difference can be found.

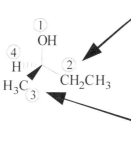

Ethyl has the higher priority because the carbon attached to the stereocenter has another carbon and two hydrogens attached (6,1,1).

Methyl has the lower priority because the carbon attached to the stereocenter has three hydrogens attached (1,1,1).

3. Multiple-bonded atoms are equivalent to the same number of singly bonded atoms.

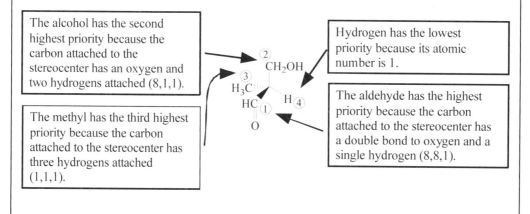

The alcohol has the second highest priority because the carbon attached to the stereocenter has an oxygen and two hydrogens attached (8,1,1).

The methyl has the third highest priority because the carbon attached to the stereocenter has three hydrogens attached (1,1,1).

Hydrogen has the lowest priority because its atomic number is 1.

The aldehyde has the highest priority because the carbon attached to the stereocenter has a double bond to oxygen and a single hydrogen (8,8,1).

Rotate the molecule so that the lowest priority group is directed **away** from you. Draw a circle from the highest priority group to the second-highest and finally the third-highest. If that circle is drawn in a *clockwise* direction, the stereocenter is designated the *R* **configuration**. If that circle is drawn in a *counterclockwise* direction, the stereocenter is designated the *S* **configuration**.

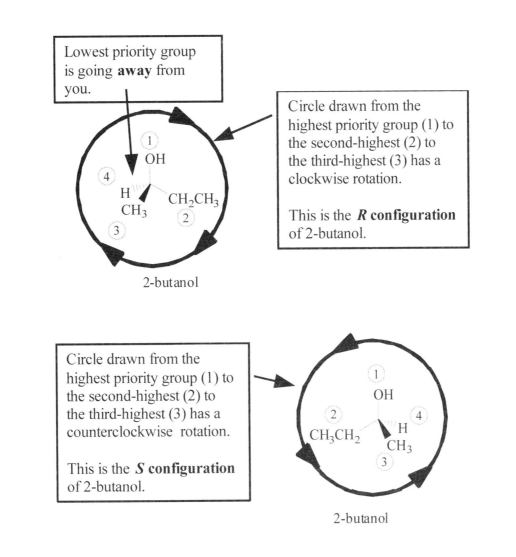

Lowest priority group is going **away** from you.

Circle drawn from the highest priority group (1) to the second-highest (2) to the third-highest (3) has a clockwise rotation.

This is the *R* **configuration** of 2-butanol.

2-butanol

Circle drawn from the highest priority group (1) to the second-highest (2) to the third-highest (3) has a counterclockwise rotation.

This is the *S* **configuration** of 2-butanol.

2-butanol

Optical Activity

A beam of light consists of electromagnetic waves. These waves can be thought of as rising and falling as they travel much like the waves at the beach. Light waves oscillate in all planes. If the beam of light is passed through a plane-polarizing filter (like that in many sun glasses or photographic lenses), then only one of the oscillation planes passes through while the others are blocked. A light wave that oscillates in a single plane is referred to as **plane-polarized light**.

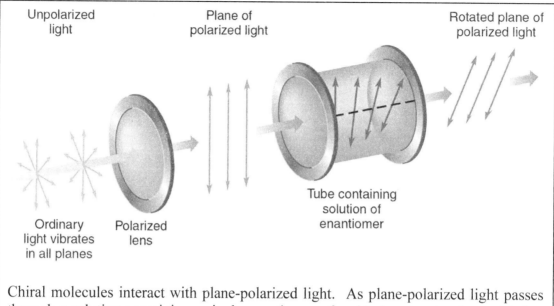

Chiral molecules interact with plane-polarized light. As plane-polarized light passes through a solution containing a single enantiomer of a chiral molecule, the oscillation plane is rotated to the left or right. One enantiomer will rotate the plane to the left by some amount and is said to be **levorotatory**. The other will rotate the beam to the right by the same amount and is said to be **dextrorotatory**.

Sample Exercises

Identifying Stereocenters

1. *How many stereocenters are found in progesterone, a hormone related to the female reproductive cycle?*

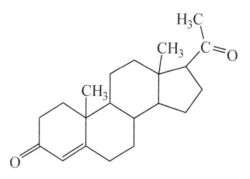

Progesterone

The correct answer is: six (6)

Remember that we're looking for tetrahedral atoms with four (4) different substituents. The stereocenters have been marked with an asterisk below.

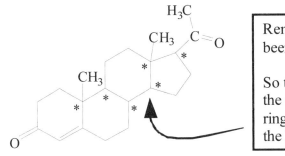

> Remember that most of the hydrogens have been omitted from the structure for clarity.
>
> So this carbon has 4 different substituents: the 6-membered ring, the 5-membered ring, the bridge between the two rings and the hydrogen.

Assigning *R* or *S* configuration

2. Assign R or S configuration to the stereocenter of adrenaline.

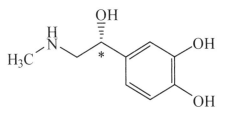

Adrenaline

The correct answer is: *R*

The first step is to prioritize the substituents attached to the stereocenter according to the *Cahn-Ingold-Prelog* rules.

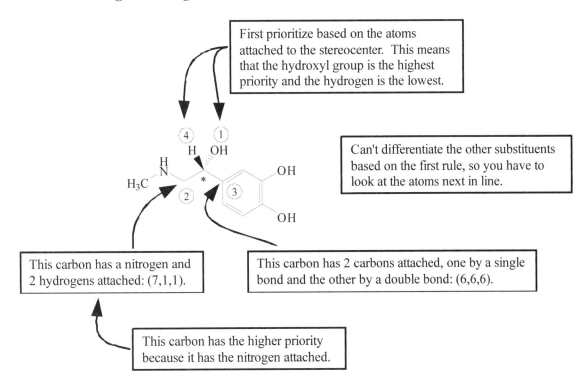

> First prioritize based on the atoms attached to the stereocenter. This means that the hydroxyl group is the highest priority and the hydrogen is the lowest.

> Can't differentiate the other substituents based on the first rule, so you have to look at the atoms next in line.

> This carbon has a nitrogen and 2 hydrogens attached: (7,1,1).

> This carbon has 2 carbons attached, one by a single bond and the other by a double bond: (6,6,6).

> This carbon has the higher priority because it has the nitrogen attached.

The next step is to rotate the molecule so that the lowest priority group is going away from you. The easiest way to do this is to keep one of the substituents in place and rotate the other three.

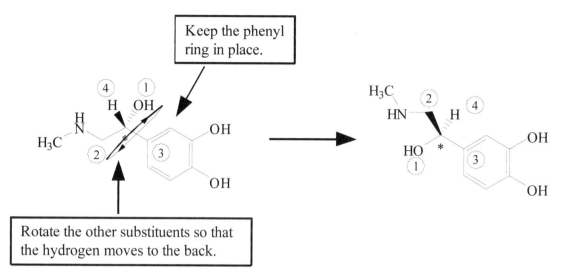

Keep the phenyl ring in place.

Rotate the other substituents so that the hydrogen moves to the back.

Now draw the circle connecting points 1, 2, and 3 and determine the configuration of the stereocenter.

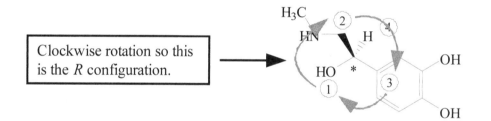

Clockwise rotation so this is the *R* configuration.

Including Stereochemistry in Compound Name
3. *Name the compound shown below.*

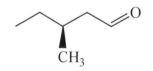

The correct answer is: (*S*)-3-methylpentanal

From Module 8, we know that this is one of the enatiomers of 3-methylpentanal. Now we must determine the configuration of the stereocenter at carbon 3.

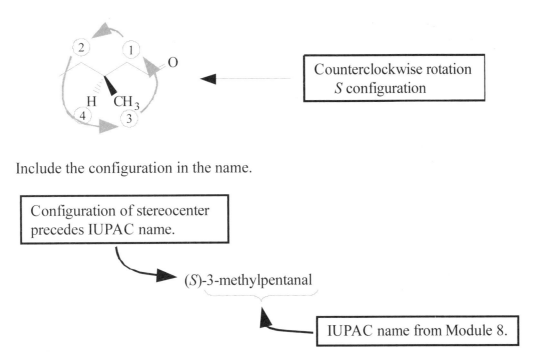

Include the configuration in the name.

Configuration of stereocenter precedes IUPAC name.

(*S*)-3-methylpentanal

IUPAC name from Module 8.

Drawing Stereoisomers
4. Draw the chemical structure of (R)-1-chloro-1-fluoroethane.

The correct answer is:

Begin by drawing the basic tetrahedral stereocenter and prioritizing the substituents.

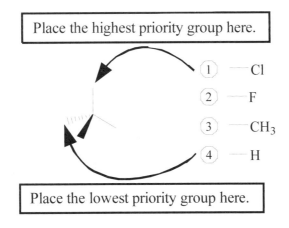

Place the highest priority group here.

① — Cl
② — F
③ — CH_3
④ — H

Place the lowest priority group here.

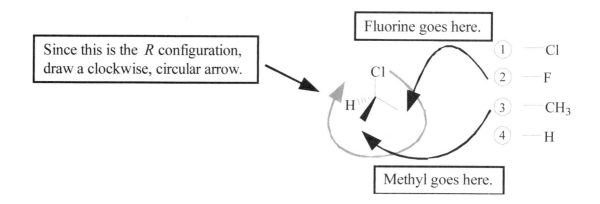

Module 10
Substitution Reactions

Introduction
Substitution reactions occur in alkanes, aromatic compounds, carboxylic acids, and the derivatives of carboxylic acids (amides, carboxylic anhydrides, and esters). Substitution reactions characteristically break one of the molecule's bonds and replace an atom or group of atoms with another atom or group. This module will help you:
1. recognize the characteristics of substitution reactions
2. predict substitution reaction products

Module 10 Key Concepts

1. **Halogenation of Alkanes**
    In halogenation reactions a halogen molecule replaces one of the hydrogens on an alkane. If the alkane reacts with additional moles of halogen, further substitution occurs. This reaction occurs in the presence of light or heat.

    $$CH_3CH_3 + Cl_2 \xrightarrow{\text{heat or light}} CH_3CH_2Cl + HCl$$

2. **Aromatic Substitutions**
    Benzene and other aromatic compounds react with halogens, nitric acid, or sulfuric acid replacing a hydrogen atom on the aromatic ring with a halogen, a nitro group (-NO$_2$), or a sulfonic acid group (-SO$_3$H), respectively.

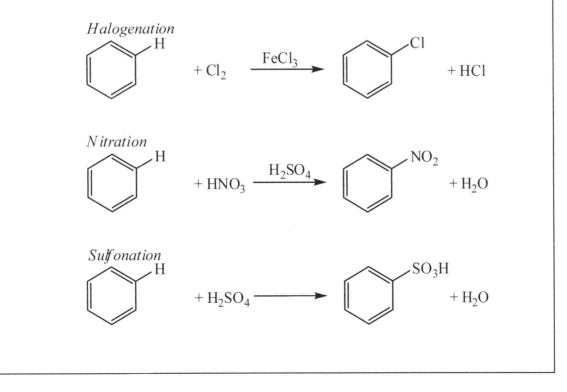

---

3. Substitutions on Carboxylic Acids and their Derivatives

Because of their differences in electronegativity carbon-oxygen double bonds in carbonyl groups are polar. Bearing the partial positive charge, the carbon atom is susceptible to attack by an electron rich species or *nucleophile*. Substitution reactions occur at the carbonyl carbons of carboxylic acids and their derivatives.

One such example is the Fischer Esterification reaction between a carboxylic acid and an alcohol to produce an ester.

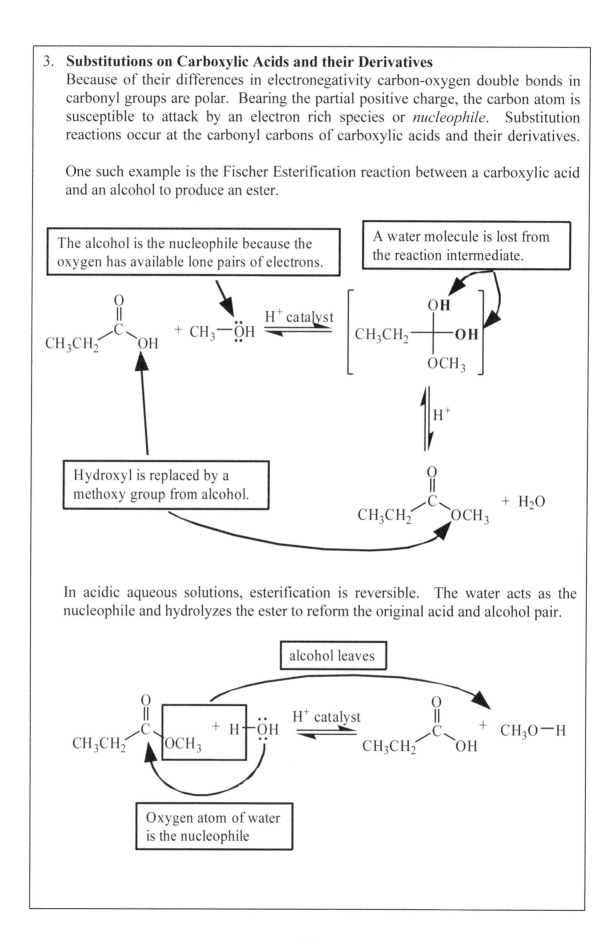

The alcohol is the nucleophile because the oxygen has available lone pairs of electrons.

A water molecule is lost from the reaction intermediate.

Hydroxyl is replaced by a methoxy group from alcohol.

In acidic aqueous solutions, esterification is reversible. The water acts as the nucleophile and hydrolyzes the ester to reform the original acid and alcohol pair.

alcohol leaves

Oxygen atom of water is the nucleophile

Like carboxylic acids, acid halides undergo substitution reactions with alcohols to yield esters and react with amines to produce amides.

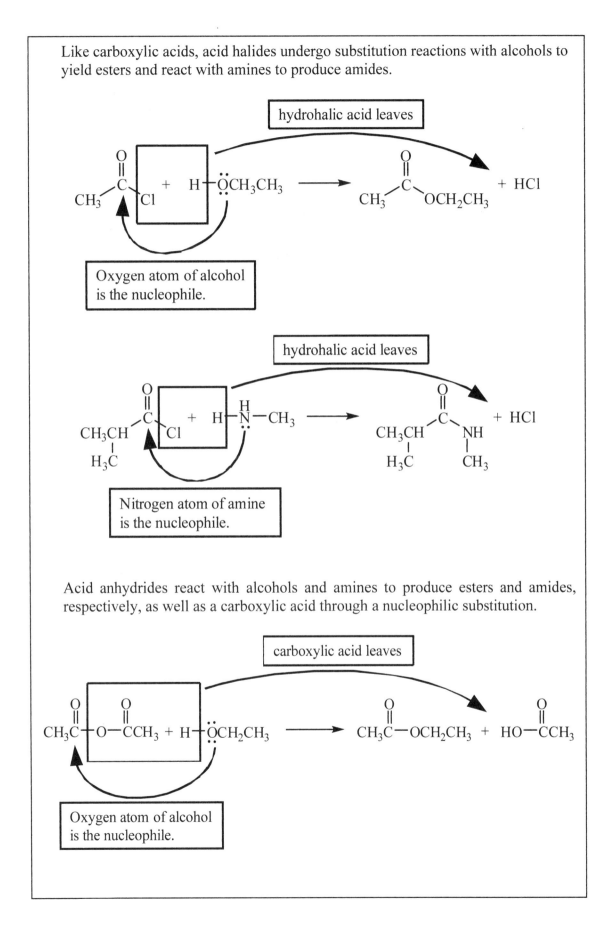

Acid anhydrides react with alcohols and amines to produce esters and amides, respectively, as well as a carboxylic acid through a nucleophilic substitution.

107

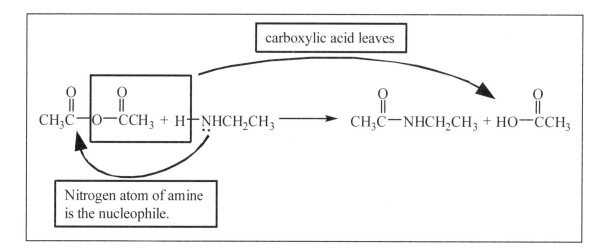

Nitrogen atom of amine is the nucleophile.

carboxylic acid leaves

Sample Exercises

Substitution Reactions of Alkanes

1. *Reaction of one mole of ethane with one mole of chlorine produces chloroethane. Further reaction with a second mole of chlorine results in two (2) dichloroethane isomers. Draw and name each of these isomers.*

The correct answer is:

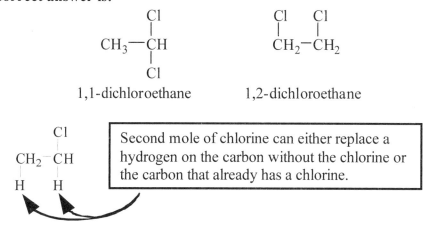

1,1-dichloroethane   1,2-dichloroethane

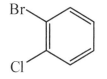

Second mole of chlorine can either replace a hydrogen on the carbon without the chlorine or the carbon that already has a chlorine.

Substitution Reactions of Benzene

2. *If one mole of benzene is first reacted with one mole of bromine and then one mole of chlorine, three products are formed. Draw and provide the IUPAC names for these compounds.*

The correct answer is:

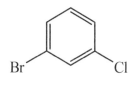

1-bromo-2-chlorobenzene

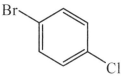

1-bromo-3-chlorobenzene   1-bromo-4-chlorobenzene

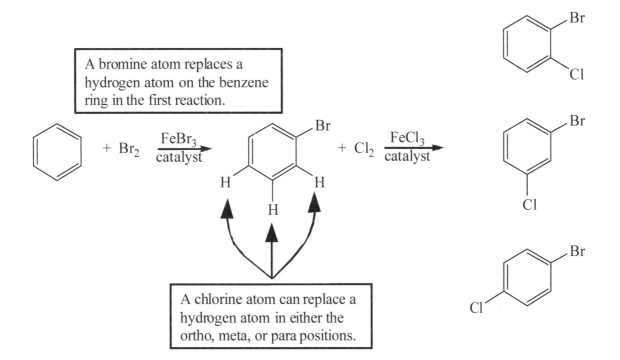

A bromine atom replaces a hydrogen atom on the benzene ring in the first reaction.

A chlorine atom can replace a hydrogen atom in either the ortho, meta, or para positions.

Substitution Reactions of Carbonyl Compounds

3. *Draw the product of the reaction of butanoic acid with phenol in the presence of an acid catalyst?*

   The correct answer is:

   phenyl butanoate

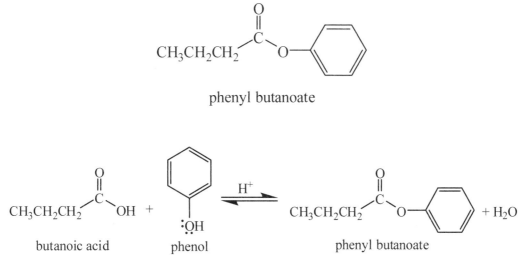

   butanoic acid   phenol   phenyl butanoate

4. *Amides and water react in the same fashion as the other nucleophilic substitutions of carboxylic acid derivatives. Identify the nucleophile and predict the products of the reaction between N-methylethanamide and water.*

The correct answer is:

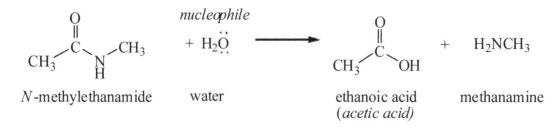

N-methylethanamide   water   ethanoic acid   methanamine
                                                    (acetic acid)

5. *What acid halide and alcohol are needed to produce ethyl 2-methylpropanoate?*

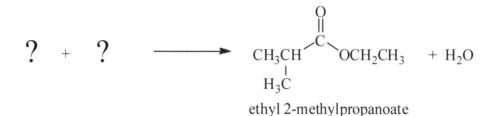

ethyl 2-methylpropanoate

The correct answers are:

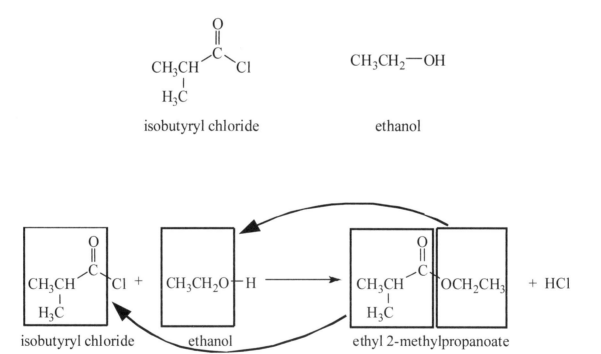

isobutyryl chloride   ethanol

isobutyryl chloride   ethanol   ethyl 2-methylpropanoate

Module 11
Addition Reactions

Introduction

Addition reactions occur at regions of high electron density on molecules, such as double and triple bonds. Alkenes, alkynes, and carbonyl groups of aldehydes and ketones undergo addition reactions with a variety of chemical species. This module will help you:

1. recognize the characteristics of addition reactions
2. understand Markovnikov's rule
3. predict addition reaction products

Module 11 Key Concepts

1. **Adding Symmetric Molecules to Alkenes**
   Hydrogen (H_2) will add to double bonds in the presence of a metal catalyst. Hydrogen addition reduces alkenes to alkanes.

$$CH_3CH_2CH{=}CHCH_3 + H_2 \xrightarrow{\ \ Pt\ \ } \overset{\overset{H}{|}}{CH_3CH_2CH}{-}\overset{\overset{H}{|}}{CHCH_3}$$

   Halogen molecules (Cl_2, Br_2, I_2) also undergo addition reactions with alkenes.

$$CH_3CH_2CH{=}CHCH_3 + Cl_2 \longrightarrow \overset{\overset{Cl}{|}}{CH_3CH_2CH}{-}\overset{\overset{Cl}{|}}{CHCH_3}$$

2. **Adding Asymmetric Molecules to Alkenes (Markovnikov's Rule)**
   Hydrohalic acids (HCl, HBr, HI) and water are example of asymmetric molecules because in addition reactions hydrogen adds to one carbon of the double bond and a halogen or hydroxyl to the other carbon. If the double bond is symmetric, meaning each double bond carbon has the same number of hydrogens attached, the hydrogen adds to either end.

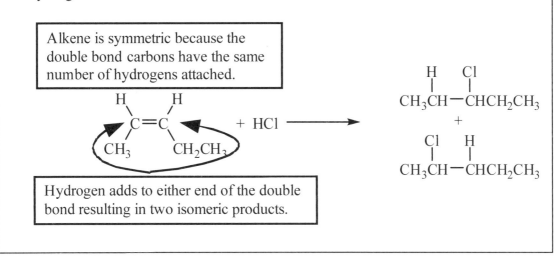

Alkene is symmetric because the double bond carbons have the same number of hydrogens attached.

Hydrogen adds to either end of the double bond resulting in two isomeric products.

111

The double bond of 2-pentene is symmetric because each carbon of the double bond has the same number of attached hydrogens. This is an important consideration when asymmetric reagents like hydrohalic acids or water add to the double bond. The hydrogen adds to either carbon of the double bond and the halide or hydroxyl adds to the other carbon.

The 2-pentene molecule is asymmetric because the double bonded carbons have two different attached alkyl groups. Because of the asymmetry of the molecule, two different isomeric products are obtained.

In general, the addition of an asymmetric reagent (HCl or H_2O) to an asymmetric molecule that contains a symmetric double bond gives two isomeric products.

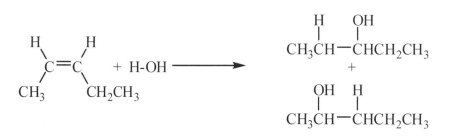

The double bond is asymmetric if the carbons of the double bond have a different number of hydrogens attached. *Markovnikov's Rule* predicts the products of addition reactions to asymmetric double bonds. When a hydrohalic acid or water adds at an asymmetric double bond, the hydrogen adds to the carbon **with the most hydrogens** and the halide or hydroxyl adds to the other carbon of the double bond.

Alkene is asymmetric because the double bond carbons have different numbers of hydrogen attached.

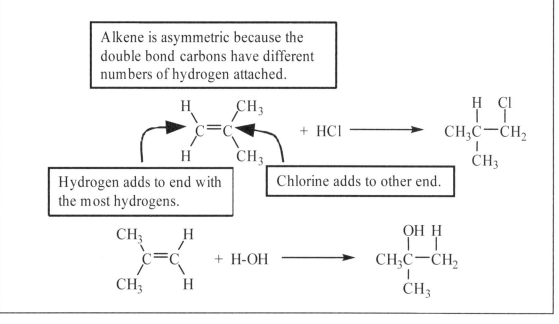

Hydrogen adds to end with the most hydrogens.

Chlorine adds to other end.

3. Nucleophilic Additions to Carbonyls

Aldehydes and ketones react with alcohols (nucleophiles because they contain the electron rich O in the hydroxyl group) by adding the alcohol to the carbonyl double bond in a mechanism similar to nucleophilic substitutions of carboxylic acids (Module 10). The addition reaction product is called a *hemiacetal*.

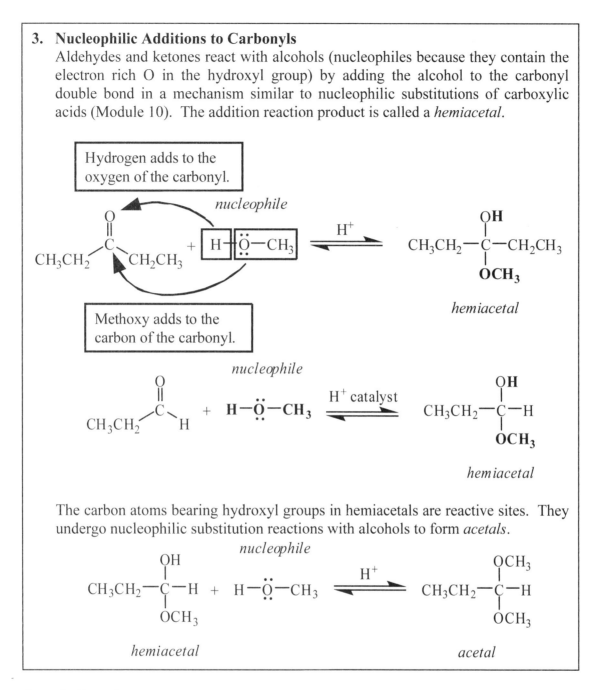

The carbon atoms bearing hydroxyl groups in hemiacetals are reactive sites. They undergo nucleophilic substitution reactions with alcohols to form *acetals*.

Sample Exercises
Addition Reactions of Alkenes

1. Write an equation for the reaction of a pentene with water in the presence of sulfuric acid that gives a single product.

The correct answer is:

$$CH_2{=}CHCH_2CH_2CH_3 + H_2O \xrightarrow[\text{catalyst}]{H_2SO_4} CH_3\overset{\underset{\displaystyle |}{OH}}{C}HCH_2CH_2CH_3$$

There are two isomers of pentene, 1-pentene and 2-pentene, so there are two possible reactions.

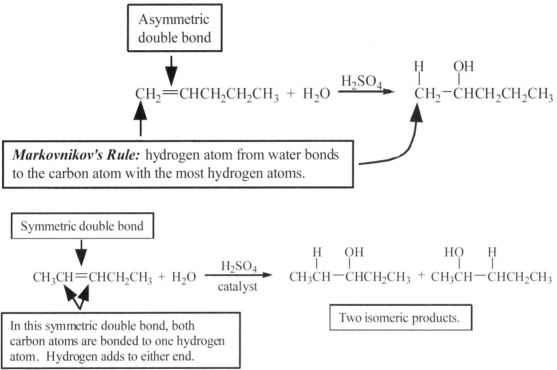

Addition Reactions of Carbonyls

2. *Most hemiacetals are unstable comprising minor components of reaction mixtures. However they are stable in molecules having both the hydroxyl (-OH) and carbonyl group which can form a five- or six-membered ring. Using this information, draw the product formed when 6-hydroxy-2-heptanone is combined with an acid catalyst?*

$$\underset{\text{6-hydroxy-2-heptanone}}{\overset{\overset{\displaystyle OH}{|}\qquad\overset{\displaystyle O}{||}}{CH_3CHCH_2CH_2CH_2CCH_3}} \underset{}{\overset{H^+}{\rightleftharpoons}} \qquad ?$$

The correct answer is:

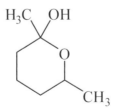

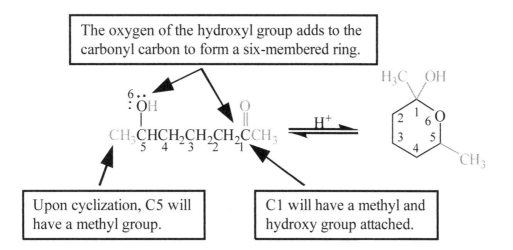

The oxygen of the hydroxyl group adds to the carbonyl carbon to form a six-membered ring.

Upon cyclization, C5 will have a methyl group.

C1 will have a methyl and hydroxy group attached.

3. **Draw the hemiacetal and acetal formed from the reaction of 5-phenyl-2-pentanone with ethanol.**

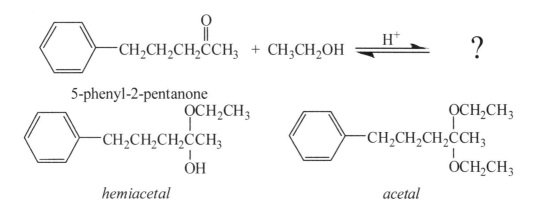

The correct answers are:

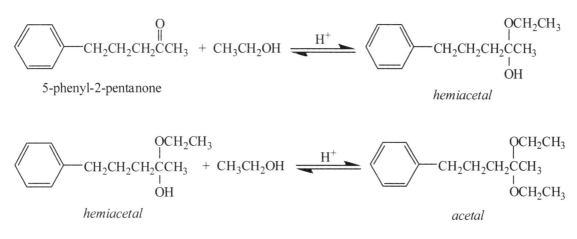

115

4. *What aldehyde or ketone and alcohol reacted to produce the acetal shown below?*

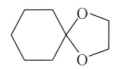

The correct answer is:

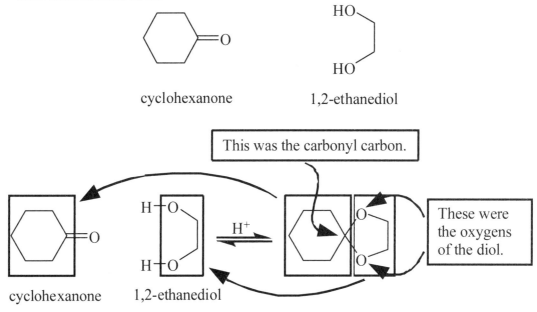

cyclohexanone   1,2-ethanediol

Module 12
Elimination Reactions

Introduction

Elimination reactions are characterized by the loss of substituents from the compound. Elimination reactions occur in alcohols and carboxylic acids. This module will help you:
1. recognize the characteristics of elimination reactions
2. predict elimination reaction products

Module 12 Key Equations & Concepts

1. **Dehydration of Alcohols**

    An alcohol can be converted to an alkene or ether through the elimination of a molecule of water. This process is called dehydration. The dehydration of alcohols is acid-catalyzed and the product depends on the starting alcohol and the temperature.

    Methanol and primary alcohols undergo an intermolecular (between two molecules) elimination of water at moderate temperatures (140°C) to form ethers.[1]

    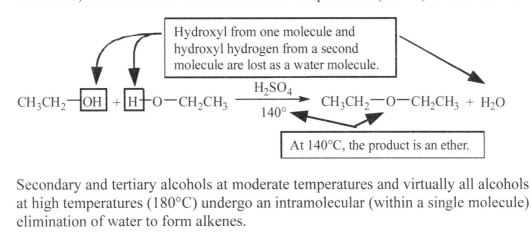

    Hydroxyl from one molecule and hydroxyl hydrogen from a second molecule are lost as a water molecule.

    At 140°C, the product is an ether.

    Secondary and tertiary alcohols at moderate temperatures and virtually all alcohols at high temperatures (180°C) undergo an intramolecular (within a single molecule) elimination of water to form alkenes.

    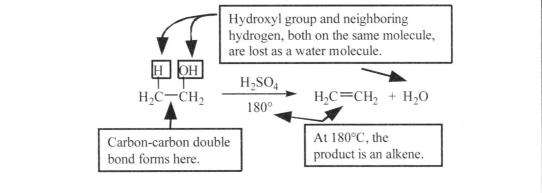

    Hydroxyl group and neighboring hydrogen, both on the same molecule, are lost as a water molecule.

    Carbon-carbon double bond forms here.

    At 180°C, the product is an alkene.

---

[1] Strictly speaking, the etherification of primary alcohols is a nucleophilic substitution, but is included here with the dehydration of alcohols to alkenes for simplicity.

If more than one isomeric alkene product is possible from the dehydration of the alcohol, the major product is the more substituted alkene.

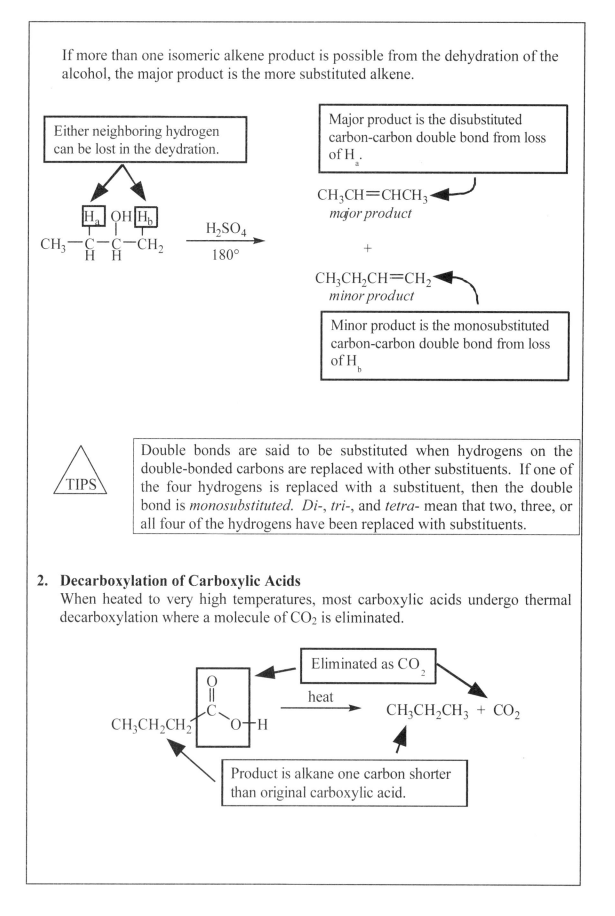

Either neighboring hydrogen can be lost in the deydration.

Major product is the disubstituted carbon-carbon double bond from loss of H_a.

$$CH_3CH = CHCH_3$$
major product

$$+$$

$$CH_3CH_2CH = CH_2$$
minor product

Minor product is the monosubstituted carbon-carbon double bond from loss of H_b

H_a OH H_b

$$CH_3 - \underset{\underset{H}{|}}{C} - \underset{\underset{H}{|}}{C} - CH_2 \xrightarrow[180°]{H_2SO_4}$$

TIPS — Double bonds are said to be substituted when hydrogens on the double-bonded carbons are replaced with other substituents. If one of the four hydrogens is replaced with a substituent, then the double bond is *monosubstituted*. *Di-*, *tri-*, and *tetra-* mean that two, three, or all four of the hydrogens have been replaced with substituents.

2. **Decarboxylation of Carboxylic Acids**
   When heated to very high temperatures, most carboxylic acids undergo thermal decarboxylation where a molecule of CO_2 is eliminated.

Eliminated as CO_2

$$CH_3CH_2CH_2 - \underset{\underset{O}{\overset{O}{||}}}{C} - O \!-\! H \xrightarrow{heat} CH_3CH_2CH_3 \ + \ CO_2$$

Product is alkane one carbon shorter than original carboxylic acid.

If there is a carbonyl group β (2 carbons away) to the carboxylic acid, the decarboxylation will occur with only moderate heating.

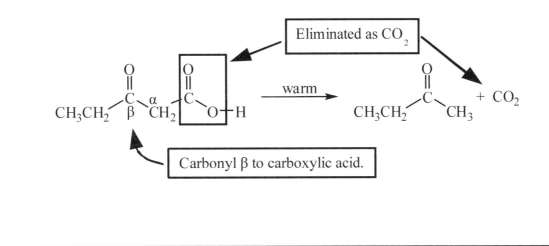

Sample Exercises
Elimination Reactions of Alcohols

1. *What is the organic product when 3-pentanol is heated to 180°C in the presence of an acid catalyst?*

$$\underset{\text{3-pentanol}}{CH_3CH_2\overset{\overset{\displaystyle OH}{|}}{C}HCH_2CH_3} \xrightarrow[180°]{H_2SO_4} \quad ?$$

The correct answer is:

$$\underset{\text{2-pentene}}{CH_3CH=CHCH_2CH_3}$$

At 180°C, the alcohol will dehydrate to form an alkene and water.

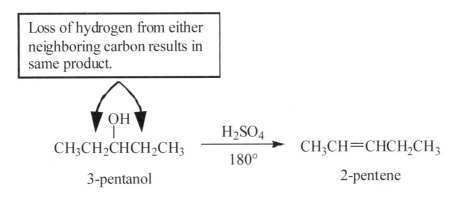

119

2. What alcohol would need to be heated to 140°C in the presence of an acid catalyst to produce 1-isobutoxy-2-methylpropane (diisobutylether)?

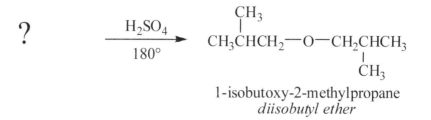

1-isobutoxy-2-methylpropane
diisobutyl ether

The correct answer is: 2-methyl-1propanol

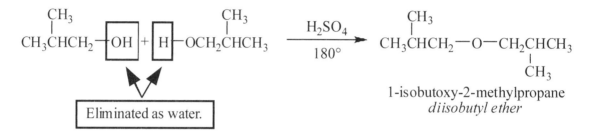

1-isobutoxy-2-methylpropane
diisobutyl ether

Eliminated as water.

3. What is the organic product formed when 2-methyl-2-propanol is heated to 140°C in the presence of an acid catalyst?

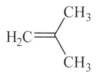

The correct answer is:

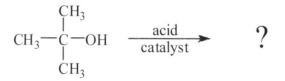

2-methylpropene

Remember that secondary and tertiary alcohols undergo intramolecular dehydration at moderate temperatures to form the alkene product, not the ether dehydration product.

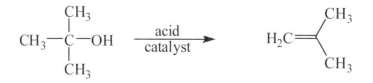

120

4. *Thermal dehydration of 2,3-dimethyl-3-pentanol yields four (4) isomeric alkene products. Draw each product, give its IUPAC name, and determine the order of products from major to minor.*

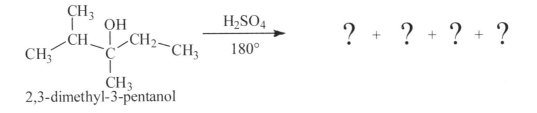

2,3-dimethyl-3-pentanol

The correct answers are:

Major Product   Minor Product

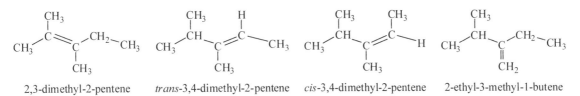

2,3-dimethyl-2-pentene   *trans*-3,4-dimethyl-2-pentene   *cis*-3,4-dimethyl-2-pentene   2-ethyl-3-methyl-1-butene

The major product will be the most substituted alkene while the minor product will be the least substituted alkene.

Major Product   **Minor Product**

tetrasubstituted alkene   *trisubstituted alkenes*   *disubstituted alkene*

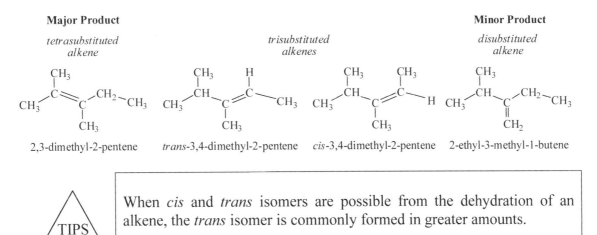

2,3-dimethyl-2-pentene   *trans*-3,4-dimethyl-2-pentene   *cis*-3,4-dimethyl-2-pentene   2-ethyl-3-methyl-1-butene

⚠ TIPS	When *cis* and *trans* isomers are possible from the dehydration of an alkene, the *trans* isomer is commonly formed in greater amounts.

Elimintation Reactions of Carboxylic Acids

5. *What carboxylic acid would need to undergo thermal decarboxylation to produce heptane?*

The correct answer is: octanoic acid

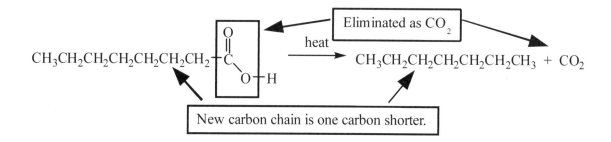

6. **What are the producst formed when 3-oxohexanoic acid is warmed?**

$$CH_3CH_2CH_2\overset{\overset{\displaystyle O}{||}}{C}CH_2\overset{\overset{\displaystyle O}{||}}{C}_{OH} \xrightarrow{\text{warm}} \qquad ?$$

3-oxohexanoic acid

The correct answer is:

$$CH_3CH_2CH_2\overset{\overset{\displaystyle O}{||}}{C}CH_3 \quad \text{and} \quad CO_2$$

2-pentanone

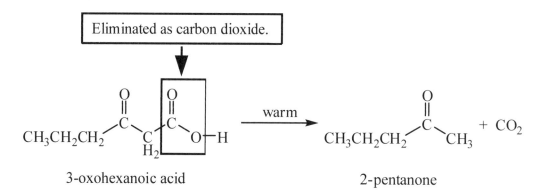

3-oxohexanoic acid   2-pentanone

122

Module 13
Reduction-Oxidation Reactions

Introduction

Reduction-Oxidation (redox) reactions involve electron exchange between chemical species. Alkanes, alcohols, aldehydes, and ketones all undergo reduction-oxidation reactions. In this module, you will learn to:

1. recognize redox reactions in organic systems
2. predict the products of redox reactions

Module 13 Key Concepts

1. **Oxidation**

   Oxidation is the process where a chemical species loses electrons. Typically in organic systems, a species is oxidized if it obtains more carbon-oxygen bonds or fewer carbon-hydrogen bonds.

2. **Oxidation of Alkanes**

   Alkanes are oxidized in combustion reactions where an alkane reacts with oxygen to produce carbon dioxide and water. In the following example, ethane is oxidized by oxygen.

   $$2\,C_2H_6 + 7\,O_2 \xrightarrow{\text{heat}} 4\,CO_2 + 6\,H_2O$$

   | Each carbon atom has 3 bonds to hydrogen and none to oxygen. | After combustion, each carbon atom has 2 double bonds to oxygen and none to hydrogen. |

   In this example, ethane is oxidized by oxygen, the oxidizing agent.

3. **Oxidation of Alcohols**

   Potassium dichromate ($K_2Cr_2O_7$), a strong oxidizing agent, oxidizes alcohols either to aldehydes, ketones, or carboxylic acids depending upon whether the alcohol is primary or secondary. Tertiary alcohols cannot be oxidized.

   a. **Primary alcohols** are rapidly oxidized by potassium dichromate to carboxylic acids. The intermediate aldehyde is oxidized by potassium dichromate to a carboxylic acid.

123

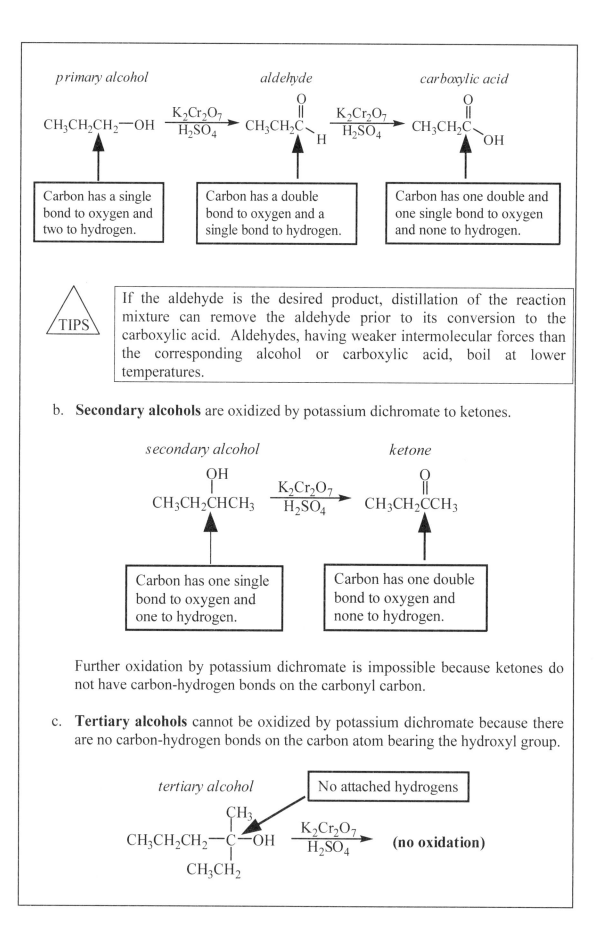

primary alcohol   *aldehyde*   *carboxylic acid*

$$CH_3CH_2CH_2{-}OH \xrightarrow[H_2SO_4]{K_2Cr_2O_7} CH_3CH_2C\overset{O}{\underset{H}{\|}} \xrightarrow[H_2SO_4]{K_2Cr_2O_7} CH_3CH_2C\overset{O}{\underset{OH}{\|}}$$

| Carbon has a single bond to oxygen and two to hydrogen. | Carbon has a double bond to oxygen and a single bond to hydrogen. | Carbon has one double and one single bond to oxygen and none to hydrogen. |

⚠️ TIPS | If the aldehyde is the desired product, distillation of the reaction mixture can remove the aldehyde prior to its conversion to the carboxylic acid. Aldehydes, having weaker intermolecular forces than the corresponding alcohol or carboxylic acid, boil at lower temperatures.

b. **Secondary alcohols** are oxidized by potassium dichromate to ketones.

secondary alcohol   *ketone*

$$CH_3CH_2\overset{OH}{\underset{}{\overset{|}{C}}}HCH_3 \xrightarrow[H_2SO_4]{K_2Cr_2O_7} CH_3CH_2\overset{O}{\underset{}{\overset{\|}{C}}}CH_3$$

| Carbon has one single bond to oxygen and one to hydrogen. | Carbon has one double bond to oxygen and none to hydrogen. |

Further oxidation by potassium dichromate is impossible because ketones do not have carbon-hydrogen bonds on the carbonyl carbon.

c. **Tertiary alcohols** cannot be oxidized by potassium dichromate because there are no carbon-hydrogen bonds on the carbon atom bearing the hydroxyl group.

tertiary alcohol | No attached hydrogens |

$$CH_3CH_2CH_2{-}\underset{CH_3CH_2}{\overset{CH_3}{\underset{|}{\overset{|}{C}}}}{-}OH \xrightarrow[H_2SO_4]{K_2Cr_2O_7} \textbf{(no oxidation)}$$

124

4. Oxidation of Aldehydes

Aldehydes are so easily oxidized that very mild oxidizing agents such as exposure to air and Tollen's reagent ($Ag(NH_3)_2Cl$) will oxidize them to carboxylic acids. Tollen's reagent oxidation is accompanied by the formation of a silver mirror on the glass walls of the test tube.

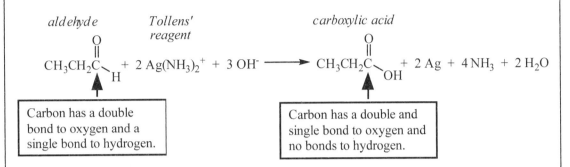

Carbon has a double bond to oxygen and a single bond to hydrogen.

Carbon has a double and single bond to oxygen and no bonds to hydrogen.

5. Oxidation of Ketones Does Not Occur

Ketones resist oxidation by potassium dichromate, Tollen's reagent and air due to the lack of carbon-hydrogen bonds on the carbonyl carbon.

ketone

$$CH_3CH_2CCH_3 \xrightarrow[H_2SO_4]{K_2Cr_2O_7} \textbf{(no oxidation)}$$

6. Reduction

Reduction occurs when a chemical species gains electrons. Reduction in organic systems is accompanied by the formation of carbon-hydrogen bonds and loss of carbon-oxygen bonds.

7. Reduction of Aldehydes and Ketones

Aldehydes and ketones are typically reduced by sodium borohydride, $NaBH_4$. Aldehydes are reduced to primary alcohols. Ketones are reduced to secondary alcohols.

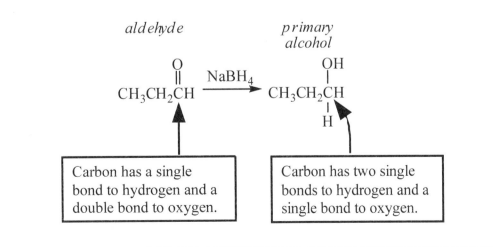

Carbon has a single bond to hydrogen and a double bond to oxygen.

Carbon has two single bonds to hydrogen and a single bond to oxygen.

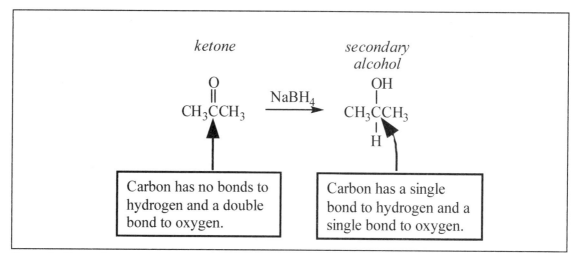

ketone

secondary alcohol

$$CH_3CCH_3 \xrightarrow{NaBH_4} CH_3CCH_3$$

Carbon has no bonds to hydrogen and a double bond to oxygen.

Carbon has a single bond to hydrogen and a single bond to oxygen.

Sample Exercises
Combustion of Alkanes
1. Write the chemical equation for the combustion of octane.

The correct answer is: $2\,C_8H_{18}\ +\ 25\,O_2\ \rightarrow\ 16\,CO_2\ +\ 18\,H_2O$

The products of the combustion of alkanes are carbon dioxide and water. Refer to Module 4, Sample Exercises 7 and 8 for help balancing chemical equations.

Oxidation Reactions of Alcohols
2. What is the organic product of the reaction between 3-pentanol and potassium dichromate?

$$CH_3CH_2CHCH_2CH_3 \xrightarrow[H_2SO_4]{K_2Cr_2O_7} \quad ?$$

OH

3-pentanol

The correct answer is:

$$CH_3CH_2CCH_2CH_3$$

O

3-pentanone

Remember that secondary alcohols are oxidized to ketones.

$$CH_3CH_2CHCH_2CH_3 \xrightarrow[H_2SO_4]{K_2Cr_2O_7} CH_3CH_2CCH_2CH_3$$

OH

O

3-pentanol
a secondary alcohol

3-pentanone
a ketone

3. *What is the product of the reaction between 2-methyl-2-butanol and potassium dichromate?*

The correct answer is: There is no reaction because 2-methyl-2-butanol is a tertiary alcohol and resists oxidation.

Reduction of Aldehydes and Ketones

4. *What is the product of the reaction between butanal and sodium borohydride?*

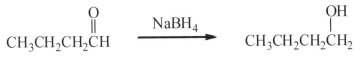

butanal

The correct answer is:

$$OH$$
$$|$$
$$CH_3CH_2CH_2CH_2$$

1-butanol

Aldehydes are reduced to primary alcohols.

$$
\begin{array}{ccc}
O & & OH \\
\| & \xrightarrow{\text{NaBH}_4} & | \\
CH_3CH_2CH_2CH & & CH_3CH_2CH_2CH_2
\end{array}
$$

butanal   1-butanol
an aldehyde   *a primary alcohol*

Module 14
Carbohydrates

Introduction
Carbohydrate is an alternative name for a sugar. Glyceraldehyde is the simplest sugar, with a molecular formula $C_3H_6O_3$. The molecular formula of glyceraldehyde can be rewritten as $C_3(H_2O)_3$ and its formula can be described as a hydrate (H_2O) of carbon (C). Molecular formulas for all sugars can be similarly rewritten as $C_n(H_2O)_n$. Simple sugars are also known as saccharides because of their sweet taste (Latin: *saccharum*, "sugar"). In this module you will learn:
1. common structural representations of sugars
2. reactions of sugars
3. polymers of sugars

Module 14 Key Concepts
1. **Fischer projections**
   Fischer projections are two-dimensional representations of molecules. The stereocenter is rotated so that the horizontal bonds from the stereocenter are projecting out of the plane towards the viewer and the vertical bonds from the stereocenter are projecting behind the plane away from the viewer. Shown below are three representations of D-glyceraldehyde for comparison.

   D-Glyceraldehyde

   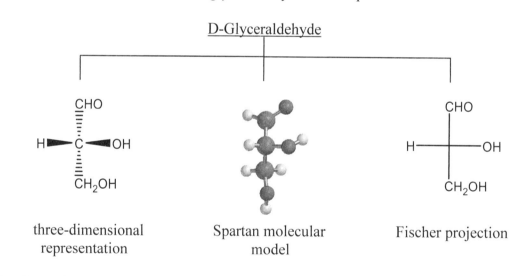

   three-dimensional representation — Spartan molecular model — Fischer projection

2. **D- and L- Nomenclature System**
   The D- and L- system was proposed by Emil Fischer in 1891. It is unique for sugars, but it is similar to the *R, S* system used to describe stereocenters in non-sugars. Glyceraldehyde possesses one stereocenter and exhibits two enantiomers. Fischer described D-glyceraldehyde as the enantiomer with the –OH group of the stereocenter on the right side of the Fischer projection. L-glyceraldehyde is the enantiomer with the –OH group of the stereocenter on the left side of the Fischer projection. The stereocenter of D-glyceraldehyde has an *R* configuration and the stereocenter of L-glyceraldehyde has an *S* configuration.

128

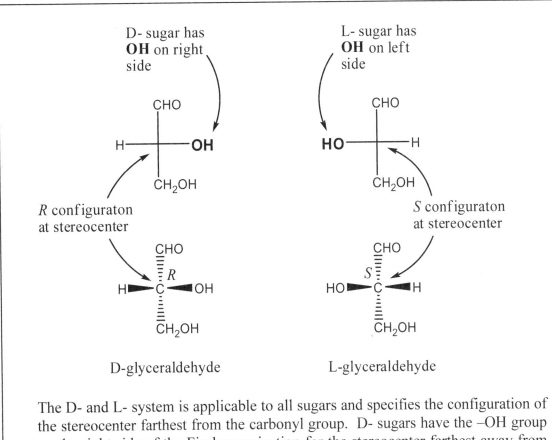

D- sugar has **OH** on right side

L- sugar has **OH** on left side

R configuraton at stereocenter

S configuraton at stereocenter

D-glyceraldehyde

L-glyceraldehyde

The D- and L- system is applicable to all sugars and specifies the configuration of the stereocenter farthest from the carbonyl group. D- sugars have the –OH group on the right side of the Fischer projection for the stereocenter farthest away from the carbonyl group, giving that stereocenter an *R* configuration. L-sugars have the –OH group on the left side of the Fischer projection for the stereocenter farthest away from the carbonyl group, giving that stereocenter an *S* configuration.

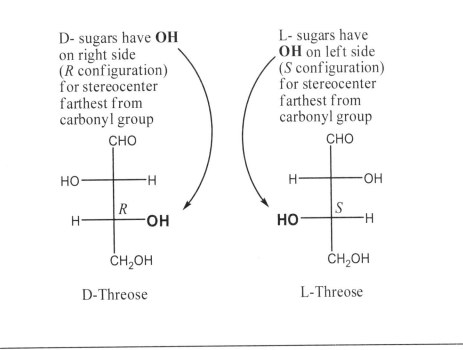

D- sugars have **OH** on right side (*R* configuration) for stereocenter farthest from carbonyl group

L- sugars have **OH** on left side (*S* configuration) for stereocenter farthest from carbonyl group

D-Threose

L-Threose

3. Cyclic sugars and Haworth projections

Sugars have hydroxyl and aldehyde or ketone groups in the same molecule and can cyclize to form an intramolecular hemiacetal. Five- and six-membered rings are common in sugars and derive their names from the five- and six-membered oxygen containing rings *furan* and *pyran*.

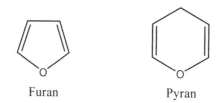

Furan   Pyran

In solution, cyclic hemiacetals exist in equilibrium with the open-chain form of the sugar. Cyclic hemiacetals interconvert by opening to the coiled form, rotating around the C1-C2 bond, and re-cyclizing, a process known as *mutarotation*. A five-membered ring (cyclic) sugar is called a *furanose* and a six-membered ring (cyclic) sugar is called a *pyranose*. When a cyclic furanose or pyranose forms, a new stereocenter is also formed. The new carbon stereocenter formed upon cyclization is called an *anomeric carbon*.

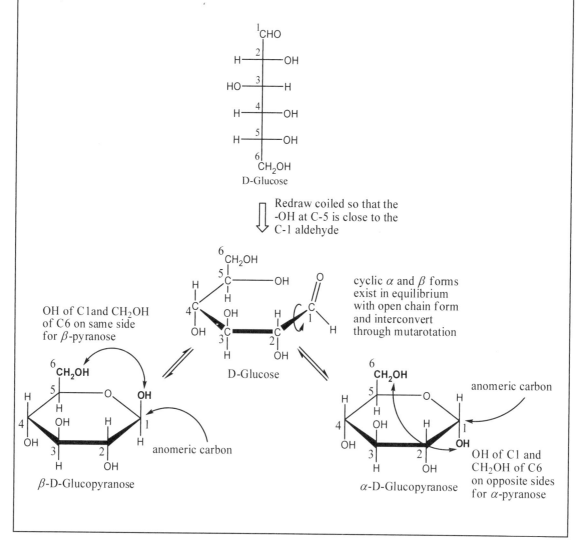

130

Haworth projections are commonly used to represent furanose and pyranose rings. In a Haworth projection, the ring is perpendicular to the plane of the paper with the anomeric carbon to the right and the hemiacetal oxygen to the back. Atoms or groups bonded to the ring carbons lie either above or below the plane of the ring. The designation β means that the –OH group on the anomeric carbon is on the same side of the ring as the terminal –CH_2OH. The designation α means that the –OH group on the anomeric carbon is on the opposite side of the ring as the terminal –CH_2OH. Recall that the most common Haworth projection has the anomeric carbon to the right and the hemiacetal oxygen to the back. In this common projection, the –CH_2OH of C6 is above the ring for D-pyranoses. Also recall that the configuration of the C5 stereocenter is R for D sugars.

Ketone containing sugars are called *ketoses*. Fructose, the most common ketose, cyclizes to form α-D-fructofuranose and β-D-fructofuranose. The C5 hydroxyl group forms a hemiacetal with the carbonyl ketone at C2. The –CH_2OH of C6 is also above the ring for D-furanoses in the common Haworth projection with the C2 carbon to the right and the C5 oxygen to the back.

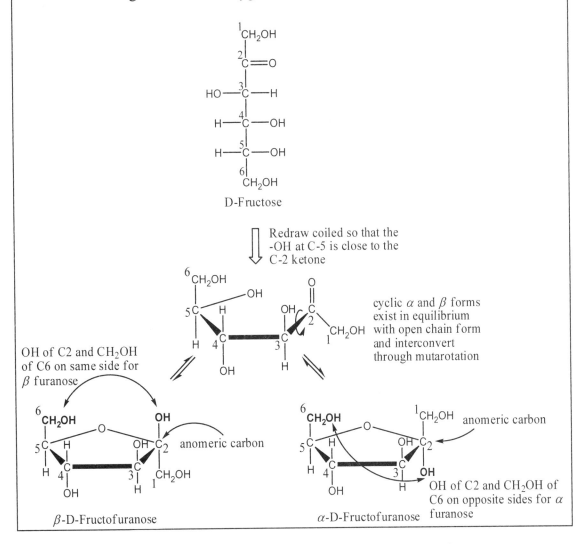

131

4. Reactions of sugars

Treatment of a furanose or pyranose with an alcohol yields an acetal. Acetals of sugars are called *glycosides* and the bond from the anomeric carbon to the –OR group is called a *glycosidic bond*. Acetals do not exist in equilibrium with an open-chain form and, consequently, glycosides do not undergo mutarotation. However, the alcohol can bond to either side of the ring as the glycoside is formed from the hemiacetal. Glycosides are named by listing the alkyl or aryl group bonded to the C1 oxygen, followed by the name of the sugar in which the ending –*e* is replaced by –*ide*.

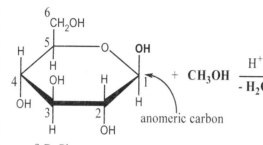

β-D-Glucopyranose

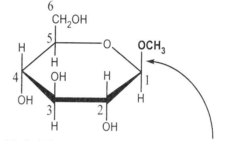

Methyl β-D-Glucopyranoside   glycosidic bond

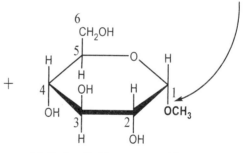

Methyl α-D-Glucopyranoside

Aldehyde containing sugars are called *aldoses*. The aldehyde group of aldoses can be selectively reduced to form alditols or oxidized to form aldonic acids.

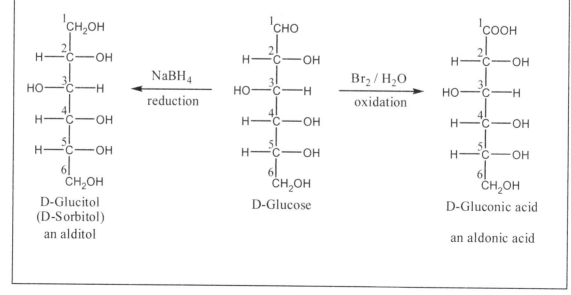

D-Glucitol
(D-Sorbitol)
an alditol

D-Glucose

D-Gluconic acid

an aldonic acid

132

5. Disaccharides and Polysaccharides

Sugars form glycosidic bonds to other sugars. This occurs when the carbonyl of one sugar molecule reacts with a hydroxyl group of another sugar molecule. Two sugars joined by a glycosidic bond is a *disaccharide*. *Sucrose* (table sugar) is the most abundant disaccharide in living systems. Sucrose contains an α-1,2-glycosidic bond from C1 of α-D-glucopyranose to C2 of β-D-fructofuranose.

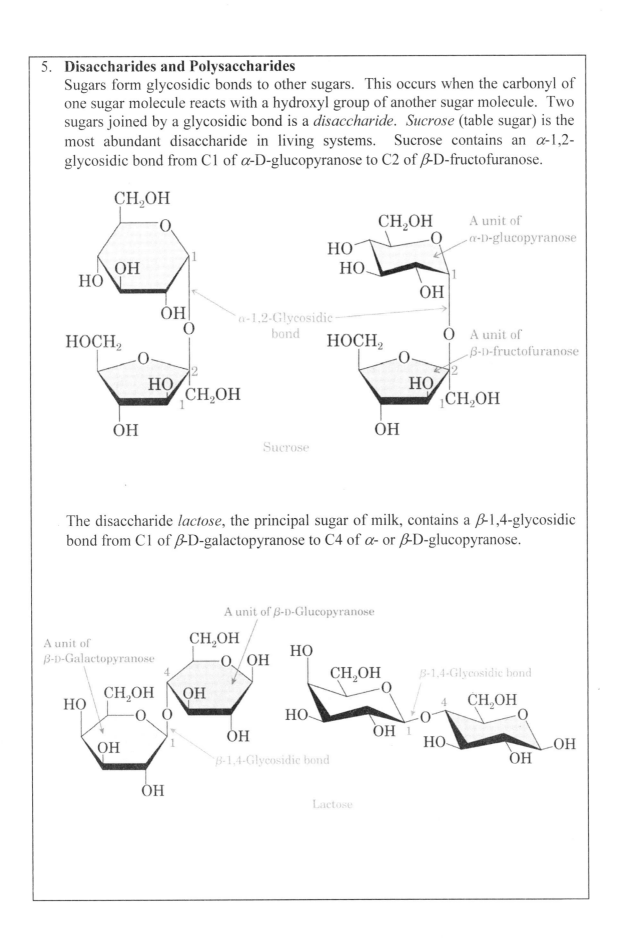

Sucrose

The disaccharide *lactose*, the principal sugar of milk, contains a β-1,4-glycosidic bond from C1 of β-D-galactopyranose to C4 of α- or β-D-glucopyranose.

Lactose

133

Maltose, the disaccharide from barley and other cereal grains, contains two D-glucopyranose units bonded by an α-1,4-glycosidic bond from C1 to C4.

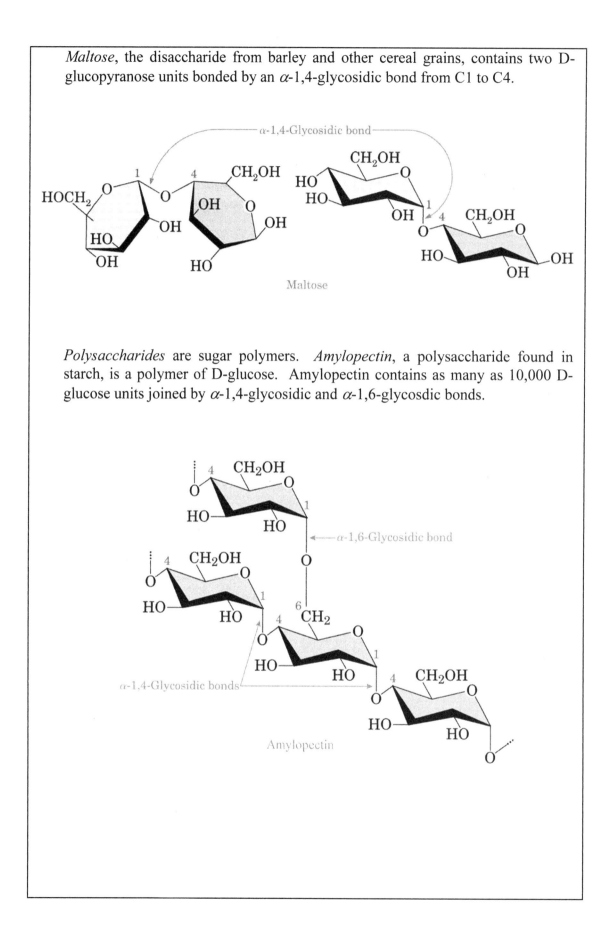

Maltose

Polysaccharides are sugar polymers. *Amylopectin*, a polysaccharide found in starch, is a polymer of D-glucose. Amylopectin contains as many as 10,000 D-glucose units joined by α-1,4-glycosidic and α-1,6-glycosdic bonds.

Amylopectin

Cellulose is the polysaccharide of wood and cotton. It is a linear combination of D-glucose units joined by β-1,4-glycosidic bonds.

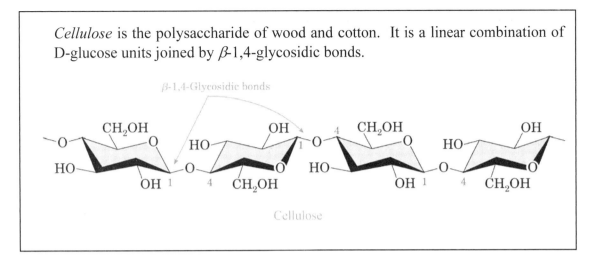

Sample Exercises

Fischer Projections

1. *Identify the following structure as a D or L form and complete the structural representation for its enantiomer.*

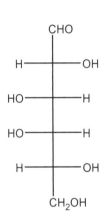

Enantiomer

The correct answer is:

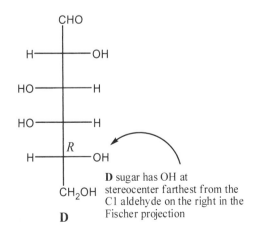

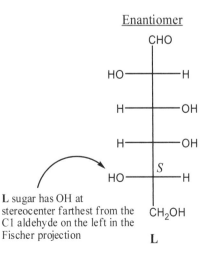

Enantiomer

The enantiomer of a given D sugar is an L sugar. The configuration at each stereocenter of the L enantiomer differs from the D sugar, including the configuration of the stereocenter farthest from the C1 aldehyde.

Haworth Projections

2. *Below are the Fischer projections for D-glucose and D-mannose and the Haworth projection for β-D-glucopyranose. Complete the Haworth projection for α-D-mannopyranose.*

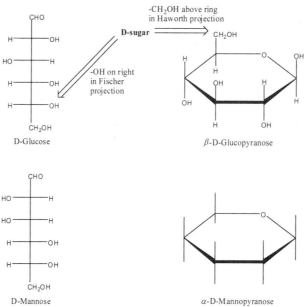

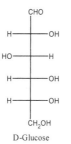

D-Mannose

α-D-Mannopyranose

The correct answer is:

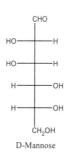

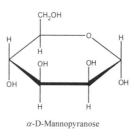

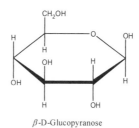

136

The Fischer and Haworth projections above are drawn in their most common representations. The Fischer projections are vertical with the aldehyde at the top. The Haworth projections are drawn with C1 to the right and the ring oxygen back. When these projections are so drawn, an –OH on the right side of a Fischer projection is below the ring in a Haworth projection. An –OH on the left in a Fischer projection is above the ring in a Haworth projection. The C5 oxygen is part of the heterocyclic ring. For the Haworth projection, the stereochemistry of C5 can be identified by the position –CH$_2$OH of C6. If –CH$_2$OH is above the ring, it is a D sugar with an R configuration at C5. Alternatively, if –CH$_2$OH is below the ring, it is an L sugar with an S configuration at C5.

3. **The Haworth projection of α-D-idose is shown below.** *(a) Draw the common Fischer projection for D-idose. (b) Draw the Haworth projection as a chair conformation.*

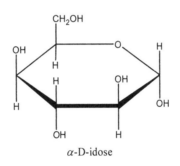

α-D-idose

The correct answers are:

(a)

(b)

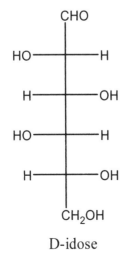

D-idose

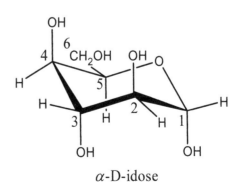

α-D-idose

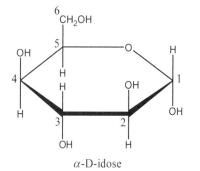

α-D-idose

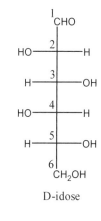

D-idose

For Haworth and Fischer projections as they are commonly drawn, an atom or group above the ring in the Haworth projection is on the left side of the Fischer projection. An atom or group below the ring in the Haworth projection is on the right side of the Fischer projection. A D sugar has the –CH₂OH of C6 above the ring in the Haworth projection. That stereochemistry is preserved when the C5 –OH is on the right in Fischer projection.

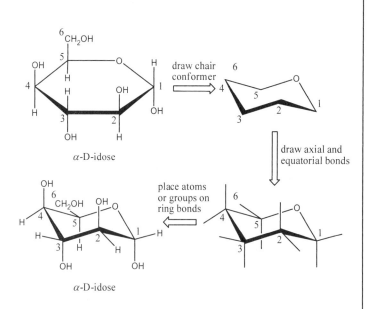

α-D-idose

draw chair conformer

draw axial and equatorial bonds

place atoms or groups on ring bonds

α-D-idose

To convert a Haworth projection to a chair conformation first draw a cyclohexane chair conformer, inserting an oxygen into the ring at the back and C1 to the right. Then, draw the axial and equatorial bonds on the ring. Finally, place the atoms or groups on the ring bonds. If an atom or group is up from the ring in the Haworth projection, place it on the bond up from the ring in the chair conformation. If the atom or group is down from the ring in the Haworth projection, place it on the bond down from the ring in the chair conformation.

138

4. *Cellobiose is a disaccharide containing a β-1-4 glycosidic bond between C1 of one D-glucopyranose unit and C4 of another D-glucopyranose unit. Draw a Haworth projection for α-cellobiose.*

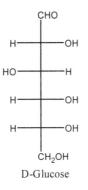

D-Glucose

The correct answer is:

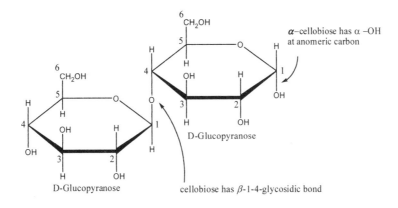

The name of the disaccharide identifies **both** monosaccharides of the disaccharide **and** the bond connecting them. *Cellobiose* is the disaccharide that contains a *β*-1-4 glycosidic bond between C1 of one D-glucopyranose unit and C4 of another D-glucopyranose unit. By comparison, *Lactose* is the disaccharide that contains an *α*-1-4 glycosidic bond between C1 of D-galactopyranose and C4 of D-glucopyranose.

α-Cellobiose has an α –OH at the anomeric carbon.

Module 15
Lipids

Introduction

Lipids are organic compounds from living organisms that are soluble in nonpolar solvents. Because they are classified on the basis of a physical property, their solubility, they are represented by a variety of compounds with different structures and biological functions. The word *lipid* comes from the Greek *lipos*, which means "fat." Lipids exhibit "fattiness" or "oiliness" because they have relatively large hydrocarbon components. In this module you will learn:

  1. structural features of lipids
  2. reactions of lipids
  3. classes and biological functions of lipids

Module 15 Key Concepts
1. **Triglycerides**

   Triglycerides, also called triacylglycerols, are triesters of glycerol and long-chain carboxylic acids called fatty acids.

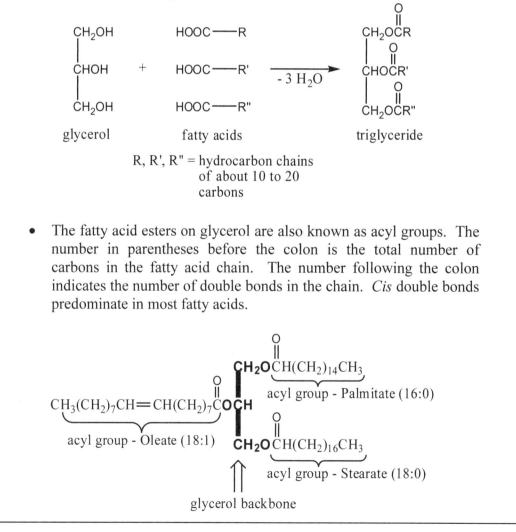

   R, R', R" = hydrocarbon chains
   of about 10 to 20
   carbons

   • The fatty acid esters on glycerol are also known as acyl groups. The number in parentheses before the colon is the total number of carbons in the fatty acid chain. The number following the colon indicates the number of double bonds in the chain. *Cis* double bonds predominate in most fatty acids.

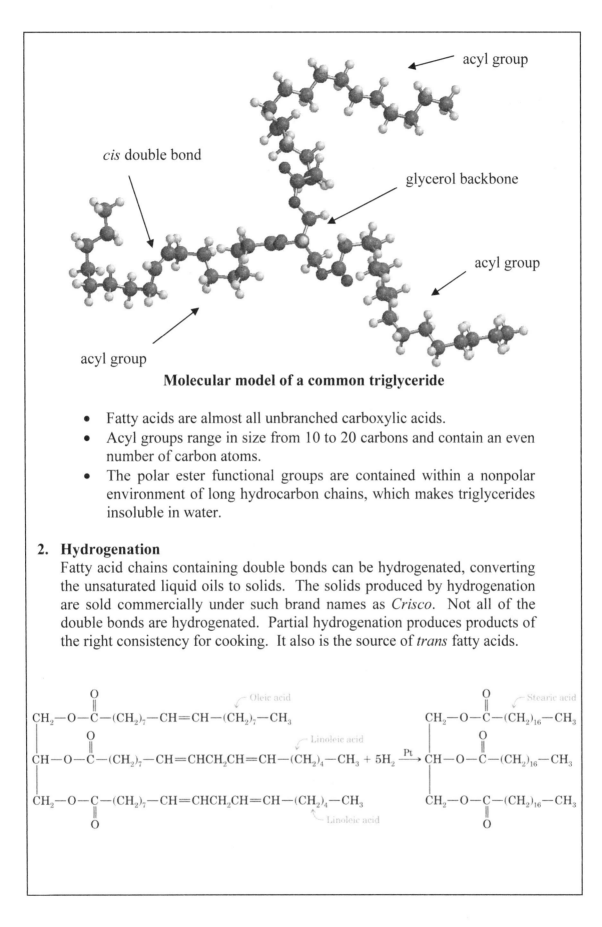

Molecular model of a common triglyceride

- Fatty acids are almost all unbranched carboxylic acids.
- Acyl groups range in size from 10 to 20 carbons and contain an even number of carbon atoms.
- The polar ester functional groups are contained within a nonpolar environment of long hydrocarbon chains, which makes triglycerides insoluble in water.

2. Hydrogenation

Fatty acid chains containing double bonds can be hydrogenated, converting the unsaturated liquid oils to solids. The solids produced by hydrogenation are sold commercially under such brand names as *Crisco*. Not all of the double bonds are hydrogenated. Partial hydrogenation produces products of the right consistency for cooking. It also is the source of *trans* fatty acids.

3. Saponification

Glycerides are esters and can be hydrolyzed in acidic or basic solution. The hydrolysis of glycerides in basic solution is known as *saponification*. The word saponification is derived from the Latin *sapon*, meaning soap. Thus, saponification is the base-promoted hydrolysis of fats and oils producing glycerol and a mixture of fatty acid carboxylates, or soaps.

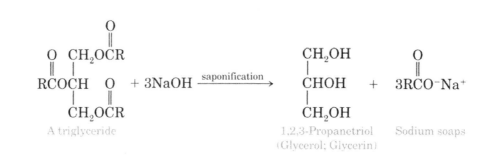

4. Complex Lipids

Complex lipids are of two general types, *phospholipids* and *glycolipids*. Phospholipids contain an alcohol, one or two fatty acids, and a phosphate group. Glycolipids contain carbohydrates and ceramides.

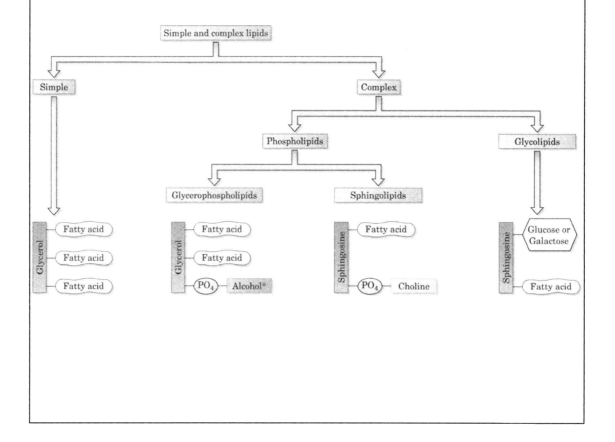

142

Complex lipids form the membranes around body cells and around small structures inside cells. Cell membranes are made of *lipid bilayers*, in which two rows (layers) of complex lipids are arranged tail to tail. The hydrophobic tails point toward each other and the hydrophilic heads project to the inner or outer surfaces of the membrane in the aqueous environment. In addition, protein molecules are suspended on the surface (peripheral proteins) or partly or fully embedded in the bilayer (integral proteins). This construction separates the aqueous environment inside cells from the aqueous environment outside cells. Transport of compounds is regulated between the inner and outer aqueous environments through the nonpolar hydrophobic tails within the membrane.

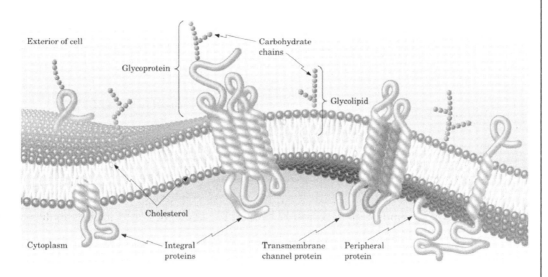

Glycerophospholipids are phospholipids with a glycerol backbone. Two of the three groups are esterified by fatty acids. The fatty acid on carbon 2 of glycerol is always unsaturated. The third group is esterified by a phosphate group, which is also esterified to another alcohol. If the other alcohol is choline, the glycerophospholipids are *phophatidylcholines*. If the other alcohol is ethanolamine or serine, the glycerophospholipids are *cephalins*. Finally, if the other alcohol is inositol, the glycerophospholipids are *phosphatidylinositols (PI)*, which are integral membrane components which also serve as signaling molecules in chemical communication.

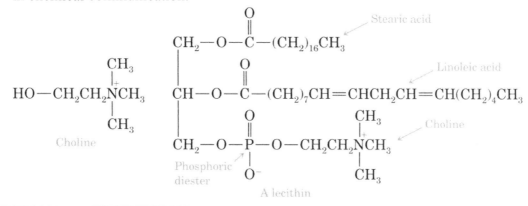

143

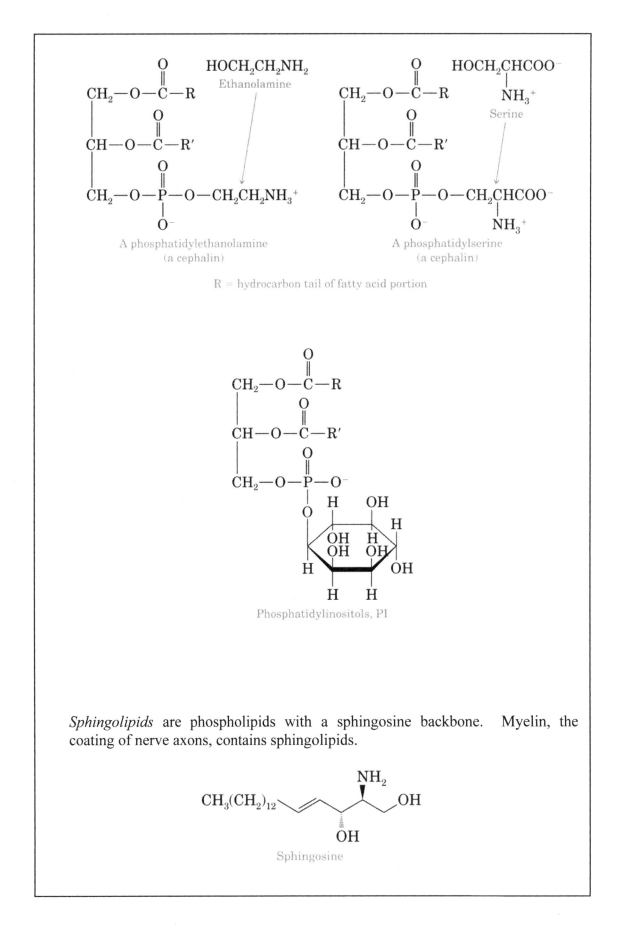

A phosphatidylethanolamine
(a cephalin)

A phosphatidylserine
(a cephalin)

R = hydrocarbon tail of fatty acid portion

Phosphatidylinositols, PI

Sphingolipids are phospholipids with a sphingosine backbone. Myelin, the coating of nerve axons, contains sphingolipids.

Sphingosine

A long-chain fatty acid is connected to the –NH₂ group by an amide bond, and the –OH group at the end of the chain is esterified by phosphorylcholine.

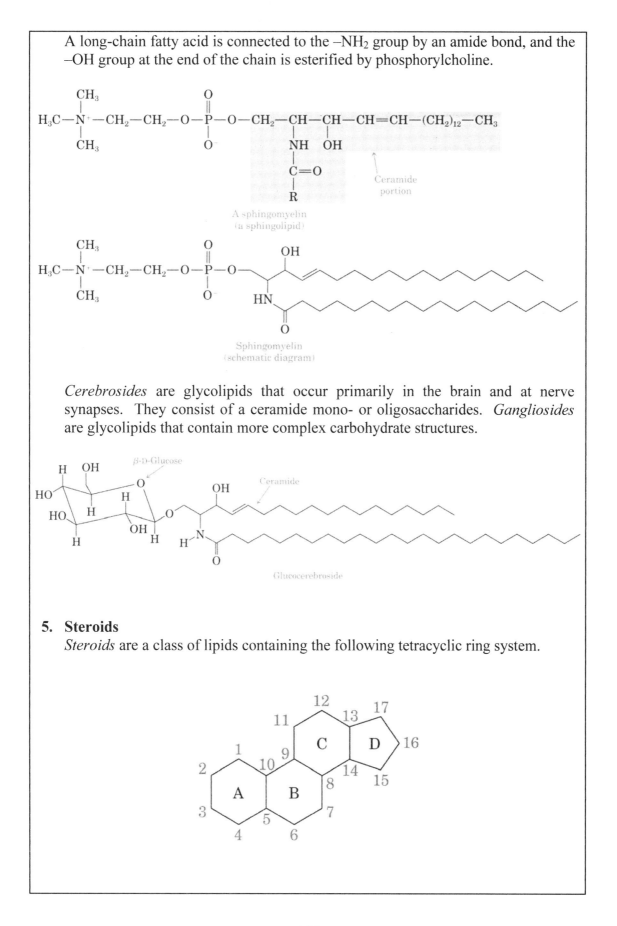

A sphingomyelin
(a sphingolipid)

Sphingomyelin
(schematic diagram)

Cerebrosides are glycolipids that occur primarily in the brain and at nerve synapses. They consist of a ceramide mono- or oligosaccharides. *Gangliosides* are glycolipids that contain more complex carbohydrate structures.

Glucocerebroside

5. Steroids

Steroids are a class of lipids containing the following tetracyclic ring system.

The most abundant steroid in the human body is *cholesterol*, which serves as a plasma membrane component in animal cells and the starting material for the biosynthesis of other steroids. Cholesterol is lipophilic. Lipoproteins solubilize and transport cholesterol in the bloodstream. *Atherosclerosis* is a narrowing of the arterial walls of blood vessels caused by plaque-like deposits of cholesterol. A low blood serum level of *Low-density lipoprotein(LDL)* accompanied by a high blood serum level of *High-density lipoprotein(HDL)* decreases the risk of atherosclerosis.

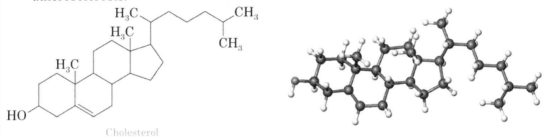

Cholesterol

Cholesterol is converted by the body into *progesterone*. Progesterone is essential for implantation of the fertilized ovum on the uterine wall, and is also the starting material for the biosynthesis of the sex and *adrenocorticoid* hormones. *Adrenal* means "adjacent to the renal (kidney)." Adrenocorticoids are classified into *mineralocorticoids, which regulate the concentrations of Na^+ and K^+ in the body,* and *glucocorticoids*, which control carbohydrate metabolism.

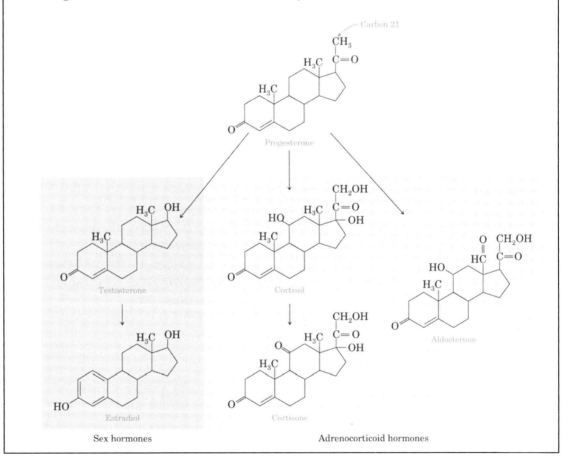

Sex hormones

Adrenocorticoid hormones

146

6. Prostaglandins and Leukotrienes

Prostaglandins, a group of lipids biosynthesized from arachidonic acid, mediate hormonal responses. They were first thought to be come from the male prostate gland, although they occur throughout the body in both sexes. Prostaglandins stimulate uterine contractions and induce labor, lower blood pressure by relaxing muscles around blood vessels, and act as decongestants. Prostaglandins are involved in the body's initial response to injury and induce platelet aggregation.

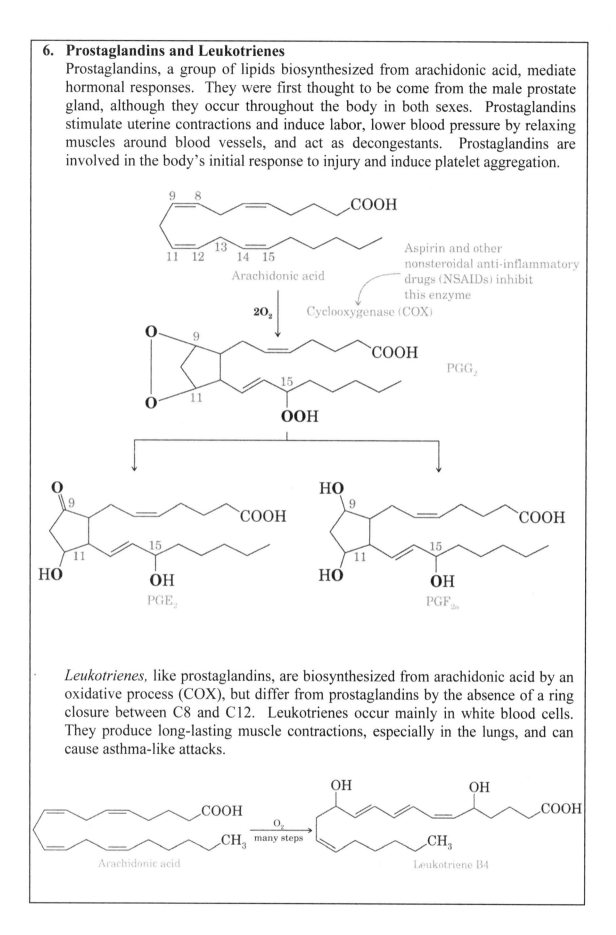

Leukotrienes, like prostaglandins, are biosynthesized from arachidonic acid by an oxidative process (COX), but differ from prostaglandins by the absence of a ring closure between C8 and C12. Leukotrienes occur mainly in white blood cells. They produce long-lasting muscle contractions, especially in the lungs, and can cause asthma-like attacks.

147

1. *Name the molecules in the box below and use them to draw structures for the*
   indicated lipids.

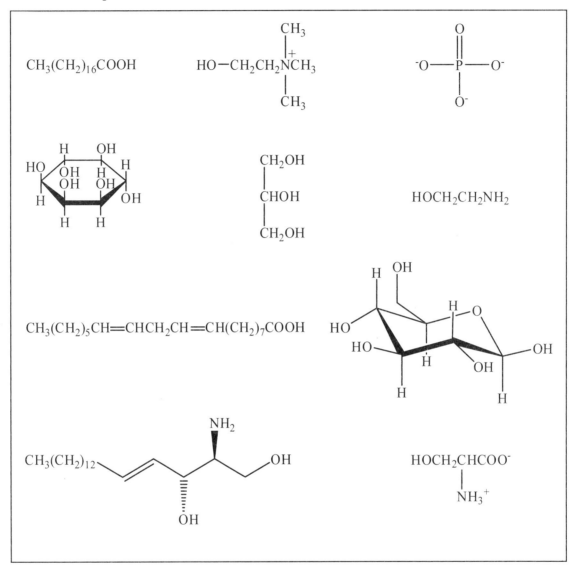

(a) a simple lipid

(b) a phosphatidylcholine

(c) a cephalin

(d) a PI

(e) a sphingolipid

(f) a glycolipid

The correct answers are:

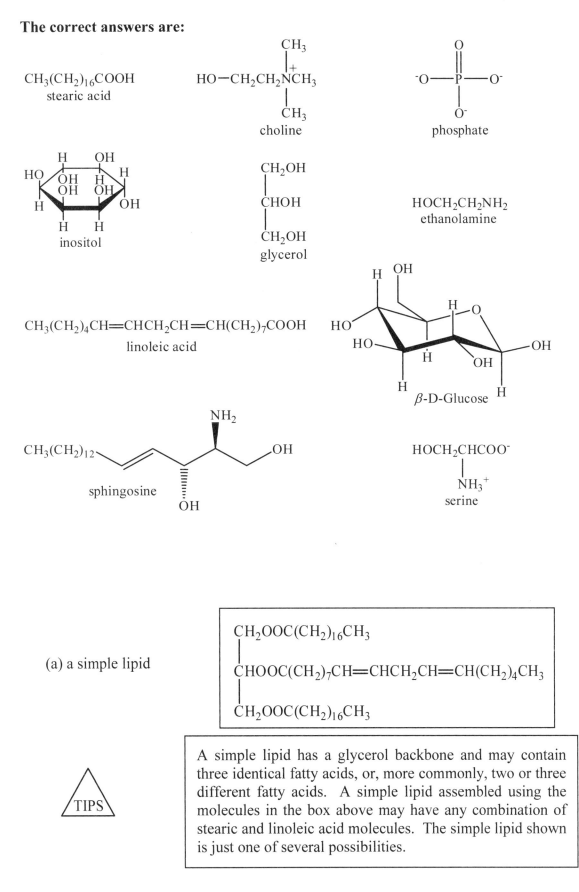

$CH_3(CH_2)_{16}COOH$
stearic acid

choline

phosphate

inositol

CH_2OH
$CHOH$
CH_2OH
glycerol

$HOCH_2CH_2NH_2$
ethanolamine

$CH_3(CH_2)_4CH{=}CHCH_2CH{=}CH(CH_2)_7COOH$
linoleic acid

β-D-Glucose

sphingosine

$CH_3(CH_2)_{12}$ ... NH_2 ... OH

OH

$HOCH_2CHCOO^-$
$NH_3{}^+$
serine

(a) a simple lipid

$CH_2OOC(CH_2)_{16}CH_3$
$CHOOC(CH_2)_7CH{=}CHCH_2CH{=}CH(CH_2)_4CH_3$
$CH_2OOC(CH_2)_{16}CH_3$

TIPS

A simple lipid has a glycerol backbone and may contain three identical fatty acids, or, more commonly, two or three different fatty acids. A simple lipid assembled using the molecules in the box above may have any combination of stearic and linoleic acid molecules. The simple lipid shown is just one of several possibilities.

(b) a phosphatidylcholine

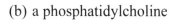

$$CH_2OOC(CH_2)_{16}CH_3$$
$$CHOOC(CH_2)_7CH=CHCH_2CH=CH(CH_2)_4CH_3$$

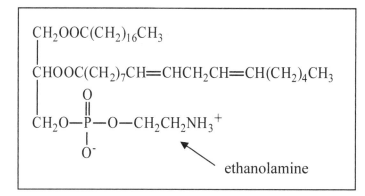

CAUTION

A phosphatidylcholine is a glycerophospholipid. All glycerophospholipids contain an unsaturated fatty acid on carbon 2 of glycerol. A phosphatidylcholine assembled from the molecules in the box will contain two linoleic acid molecules, or stearic acid on the terminal carbon and linoleic acid on carbon 2.

(c) a cephalin

$$CH_2OOC(CH_2)_{16}CH_3$$
$$CHOOC(CH_2)_7CH=CHCH_2CH=CH(CH_2)_4CH_3$$

ethanolamine

TIPS

A cephalin is a glycerophospholipid and must contain an unsaturated fatty acid on carbon 2. The other alcohol of the phosphate ester may be ethanolamine (shown above) or serine.

(d) a PI

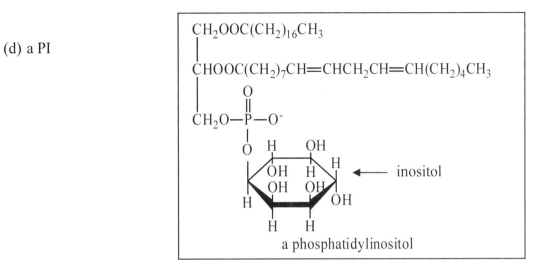

CH$_2$OOC(CH$_2$)$_{16}$CH$_3$

CHOOC(CH$_2$)$_7$CH=CHCH$_2$CH=CH(CH$_2$)$_4$CH$_3$

CH$_2$O—P—O$^-$

H   OH

OH   H   H   ← inositol

OH   OH

H   OH

H   H

a phosphatidylinositol

(e) a sphingolipid

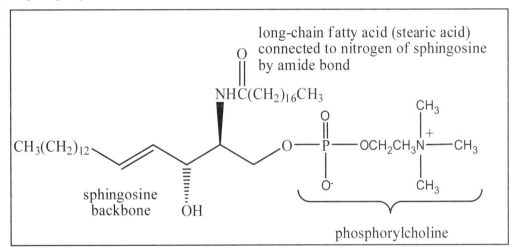

long-chain fatty acid (stearic acid) connected to nitrogen of sphingosine by amide bond

NHC(CH$_2$)$_{16}$CH$_3$

CH$_3$

CH$_3$(CH$_2$)$_{12}$ O—P—OCH$_2$CH$_2$N$^+$—CH$_3$

O$^-$ CH$_3$

sphingosine backbone   OH

phosphorylcholine

(f) a glycolipid

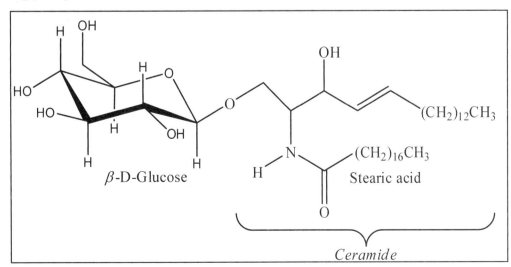

H   OH

HO

HO

H

H   OH

H   H

β-D-Glucose

OH

(CH$_2$)$_{12}$CH$_3$

N   (CH$_2$)$_{16}$CH$_3$

H   Stearic acid

O

Ceramide

151

The combination of a fatty acid (stearic acid in the above example) and sphingosine is called the *ceramide* portion of the sphingolipid, so named because many of these compounds are found in cerebrosides.

2. **Beginning with cholesterol, perform the indicated modifications and identify the hormones that are produced.**

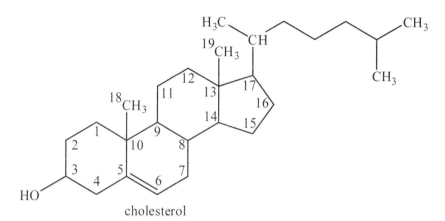

cholesterol

(a) Move (isomerize) the C5-C6 double bond to C4-C5, change (oxidize) the C3 – OH to a carbonyl, and remove the branched-chain hydrocarbon at C17 and replace it with an acetyl group (COCH₃).

(b) Beginning with the product in (a), introduce a double-bonded oxygen at C11, introduce an –OH group at C17, and change the CH₃ of the acetyl group to CH₂OH.

The correct answers are:

(a) progesterone

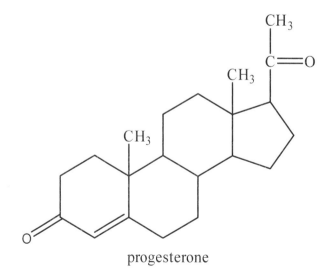

progesterone

152

(b) cortisone

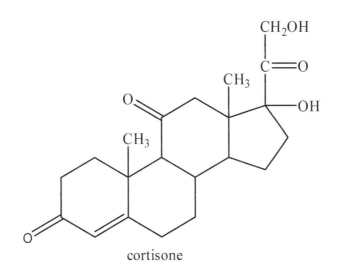

cortisone

Module 16
Proteins and Enzymes

Introduction
Every external feature of your body is composed of proteins. Proteins constitute your muscles, tendons, and many other internal structural components of humans. Proteins are also the principle components of biological catalysts called enzymes. Enzymes are such effective catalysts, that without them, many of the necessary chemical reactions for life would occur too slowly to sustain life. This module will help you understand the:
1. chemical structure of proteins
2. primary, secondary, tertiary, and quaternary structures of proteins
3. chemical structure of enzymes
4. functions of enzymes

Module 16 Key Concepts
1. **Amino Acids**

    Proteins are polymers composed of amino acids. All amino acids contain the amine and carboxyl functional groups. 20 amino acids are commonly found in proteins. Two amino acids are shown below.

    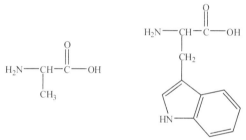

    alanine   tryptophan

2. **Structural features of proteins**

    Proteins can be quite large and complex molecules. Consequently, their structure is very complex. Their structural classification is subdivided into four categories.
    a. 1^0 (primary) structure – amino acid sequence in the protein chain
    b. 2^0 (secondary) structure – folding and alignment patterns of protein chains The four known patterns are:
       i. α - helix
       ii. β - pleated sheet
       iii. random coil
       iv. extended helix
    c. 3^0 (tertiary) structure – three dimensional structure of every atom in a protein including side chain interactions
    d. 4^0 (quaternary) structure – three dimensional structure and interactions between various subunits in proteins having more than one protein chain

3. **Enzymes are highly specific biocatalysts**

    Most enzymes are so specific in their function that they affect only a few, frequently just one, reactions. But they can enhance reaction rates by many orders of magnitude, commonly 10^9 to 10^{20} times faster.

Sample Exercises
Peptide Bonds

1. In the formation of a protein from amino acids, what functional group is formed in the reaction of amino acids to make proteins?

 a. carboxyl b. amine c. ether d. amide e. ester

The correct answer is d. amide.

An amide functional group (peptide linkage in proteins) consists of a carbonyl and an amine group bonded together as shown below.

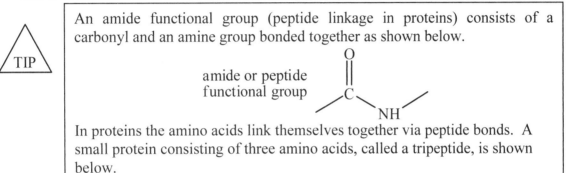

amide or peptide functional group

In proteins the amino acids link themselves together via peptide bonds. A small protein consisting of three amino acids, called a tripeptide, is shown below.

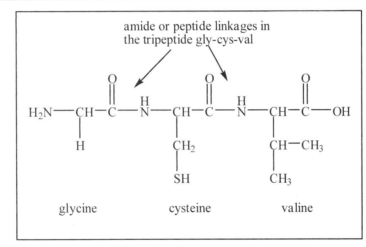

amide or peptide linkages in the tripeptide gly-cys-val

glycine   cysteine   valine

Abbreviated Peptide Names

2. Shown below is a tripeptide made from the three amino acids glycine, cysteine, and valine. Choose the correct 3-letter abbreviation name for this tripeptide.

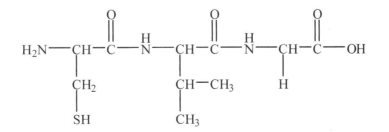

155

a. cys-val-gly   *b. cys-gly-val*   *c. gly-val-cys*

d. gly-cys-val   *e. val-cys-gly*

The correct answer is a. cys-val-gly.

 Amino acid ordering in peptides and proteins is crucial to their structure. Six different tripeptides can be made from the amino acids cysteine, valine, and glycine. Each one is unique and therefore uniquely named using a 3-letter abbreviation. These are read starting with the amine group on the left and finishing with the carboxyl group on the right. You must learn how to decipher these 3-letter abbreviations and recognize the correct structures.

3. *Draw the structures of the two tripeptides val-cys-gly and gly-cys-val.*
   The correct answers are:

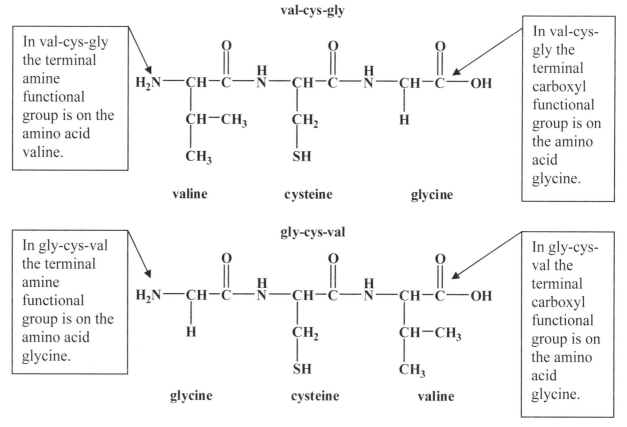

In val-cys-gly the terminal amine functional group is on the amino acid valine.

In val-cys-gly the terminal carboxyl functional group is on the amino acid glycine.

In gly-cys-val the terminal amine functional group is on the amino acid glycine.

In gly-cys-val the terminal carboxyl functional group is on the amino acid glycine.

 Placement of the terminal amine and carboxyl groups is indicates where to begin the naming system. Compare the structures of both val-cys-gly and gly-cys-val with that of cys-val-gly from Sample Exercise 2 paying attention to the terminal functional groups.

Primary Structure of Proteins

4. *What important role do the S atoms in cysteine play in the protein structure?*

The correct answer is that cysteine units from one amino acid chain bond to cysteine units in other chains with disulfide bonds. This connects the chains. Interchain disulfide linkages create bends in the proteins' 2^0 structure.

Intrachain disulfide linkage which bends the peptide chain.

Interchain to chain disulfide bonds in bovine insulin

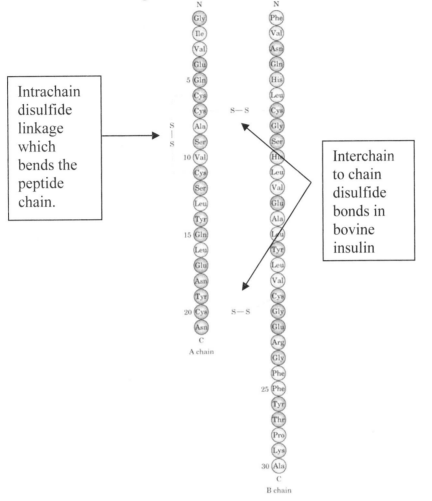

TIP

Primary structure in proteins refers to the precise order of the amino acids in the peptide chain. In the figure above there are 21 amino acids in the A chain and 30 in the B chain. Each chain has a unique ordering of the amino acids plus the two chains must correctly connect at the cysteine positions. Misplacement of a single amino acid out of the 51 in this protein would generate another protein whose biochemical activity would be different from this one. You must understand the important role of 1^0 structure in proteins.

Secondary Structure of Proteins
5. *Which of these are secondary protein structures?*
    a. double helix   b. β-pleated sheet   c. octahedral
    d. α-helix   e. prismatic
    The correct answer is b. β-pleated sheet and d. α-helix.

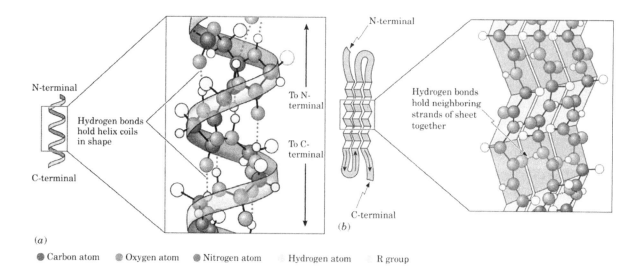

N-terminal

Hydrogen bonds
hold helix coils
in shape

C-terminal

To N-terminal

To C-terminal

(a)

N-terminal

Hydrogen bonds
hold neighboring
strands of sheet
together

C-terminal

(b)

● Carbon atom　　● Oxygen atom　　● Nitrogen atom　　Hydrogen atom　　R group

---

◇ **CAUTION**

Protein function is highly shape dependent. To understand their functionality, we must appreciate how the proteins fold themselves in three dimensions exposing certain functional groups to the exterior environment while hiding others in their interiors. Large proteins usually have regions of the molecule that contain α- helices while other regions have β-pleated sheets or random coils. You must understand these fundamental shapes to understand their biochemistry.

Tertiary Structure of Proteins

6. Which of these chemical forces stabilizes the tertiary structures of proteins?
    a. London dispersion forces　　*b. covalent bonds*　　*c. hydrogen bonds*
    d. metal ion coordination　　　*e. dipole-dipole attractions*

The correct answer is b. covalent bonds, c. hydrogen bonds, and d. metal ion coordination. Salt bridges and hydrophobic interactions are also employed.

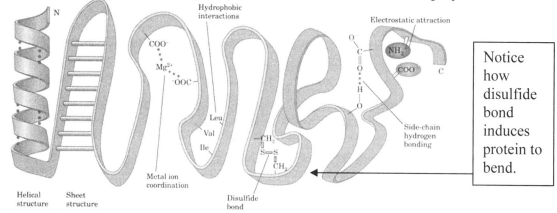

Notice how disulfide bond induces protein to bend.

Quaternary Structure of Proteins

7. How many heme molecules are present in the quaternary structure of hemoglobin?
    a. 1　　　　*b. 2*　　　　*c. 3*　　　　*d. 4*　　　　*e. 5*

The correct answer is d. 4.

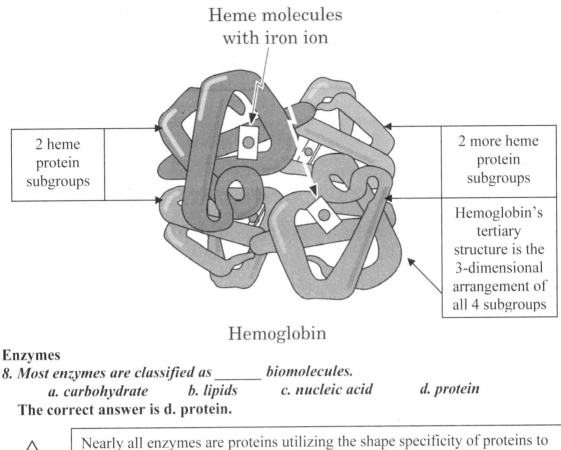

Heme molecules
with iron ion

| 2 heme protein subgroups | | 2 more heme protein subgroups |

Hemoglobin's tertiary structure is the 3-dimensional arrangement of all 4 subgroups

Hemoglobin

Enzymes

8. Most enzymes are classified as _____ biomolecules.

a. carbohydrate   *b. lipids*   *c. nucleic acid*   *d. protein*

The correct answer is d. protein.

> Nearly all enzymes are proteins utilizing the shape specificity of proteins to enhance reaction rates by many orders of magnitude. Some enzymes are so specific that they break only one type of bond but they can do so in milli- and microseconds.
>
> TIP

Enzyme Nomenclature

9. Which enzyme class catalyzes the bond formation between two molecules?

a. oxidoreductases   *b. hydrolases*   *c. lyases*

d. isomerases   *e. ligases*

The correct answer is e. ligases.

> One classification scheme for enzymes is based upon their chemical reactivity.
>
> a) Oxidoreductases catalyze oxidations and reductions
> b) Transferases catalyze atom group transfer from one molecule to another.
> c) Hydrolases catalyze hydrolysis reactions.
> d) Lyases catalyze addition or removal reactions at double bonds.
> e) Isomerases catalyze isomerization reactions.
> f) Ligases, sometimes called synthetases, catalyze bond formation between two molecules.
>
> TIPS

Models of Enzyme Activity

10. How does the lock-and-key model of enzyme activity differ from the induced-fit model?

The correct answer is that in the lock-and-key model the active site molecule retains its shape before and after the enzyme interacts with the molecule. In the induced-fit model the active site molecule's shape changes after the enzyme interacts with it to accommodate the substrate molecule.

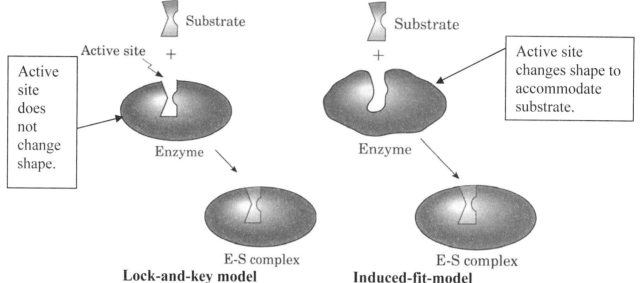

11. In enzymes changing the pH often affects the reaction rate. From a structural perspective explain why this occurs.

The correct answer is that enzymes are proteins having a specific shape which affects their reactivity. If the pH is too acidic, the enzyme denatures, altering the necessary shape to perform its function. A solution that is too basic also changes the enzyme's shape and function. However, in a narrow pH band the enzyme has the correct shape to catalyze biochemical reactions and perform as an enzyme.

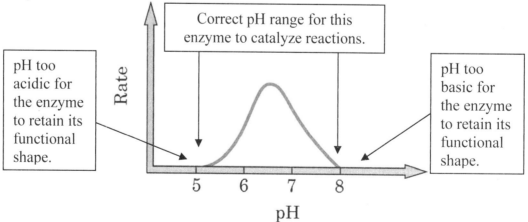

12. Choose the condition which does _not_ regulate enzymes.

    a. feedback control   **b. catabolism**   **c. allosterism**
    d. protein modification   **e. protein modification**

The correct answer is b. catabolism. Enzymes are regulated by five different mechanisms a) feedback control, b) proenzymes, c) allosterism, d) protein modification, and e) isoenzymes.

It is a good idea to know not only the names of these enzyme regulation processes, but also to understand each one's mechanism. Detailed descriptions of each process are given in your textbook.

Module 17
Nucleotides and Nucleic Acids

Introduction

Chemical molecules that form the basis of heredity are called nucleotides. They are the chemical compounds which constitute DNA and RNA. This module will help you understand the:

1. chemical structure of nucleotides
2. chemical structure of DNA
3. primary and secondary structures of DNA.

Module 17 Key Concepts

1. **Purines and Pyrimidines**

   Nucleotides are bases derived from two polycyclic aromatic bases named purine and pyrimidine.

   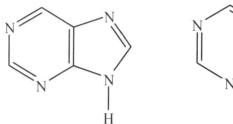

           Purine   Pyrimidine

2. **Ribose and Deoxyribose**

   Nucleic acids are composed of nucleotides, phosphate groups, and one of the two sugars β-D-Ribose or β-2-Deoxy-D-ribose.

   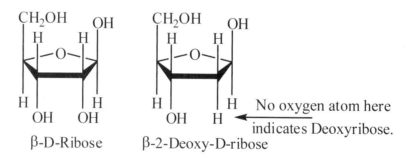

3. **Hydrogen bonding occurs between nucleotides in DNA**

   Hydrogen bonding between complementary base pairs in DNA is an important source of structural stabilization in the secondary structure of DNA.

4. **Nucleic acids have multiple layers of structure**

   Both RNA and DNA have a primary structure consisting of a phosphate, sugar, and nucleotide combination; for DNA the secondary structure is a double helix; in RNA the secondary structure depends upon the RNA function; and both nucleic acids have multiple tertiary structures.

Purines and Pyrimidines
1. Choose the aromatic heterocyclic amines that are pyrimidine derivatives.

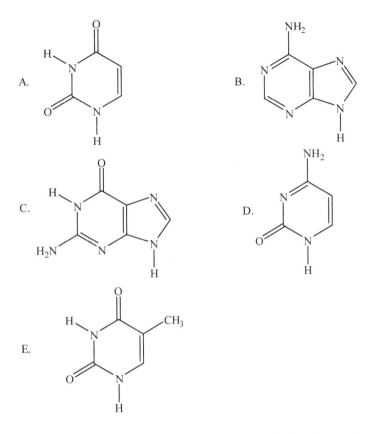

A.

B.

C.

D.

E.

The correct answer is compounds A, D, and E are derivatives of pyrimidine. Compounds B and C are purine derivatives.

TIP

To identify pyrimidine based bases look for a single 6-membered polycyclic aromatic ring. Purine derivatives have two polycyclic rings one 5-membered and one 6-membered that share a common face.

Identifying aromatic heterocyclic amines (bases in DNA and RNA)
2. Which compound is guanine?

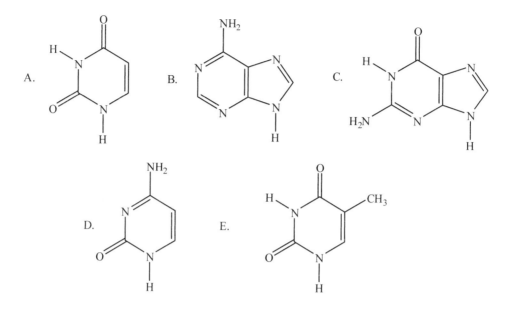

The correct answer is compound C is guanine.

Shown below are the chemical structures and names of the five aromatic heterocyclic amines present in DNA and RNA. These structures are commonly used in descriptions of DNA and RNA. You should be familiar with them.

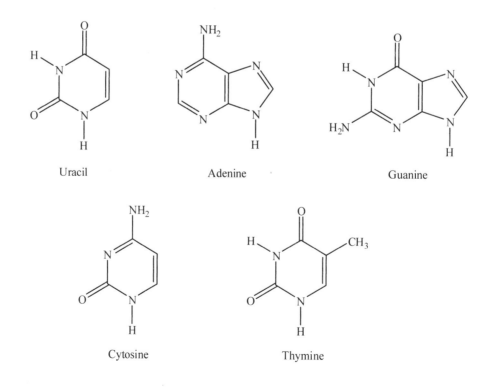

Uracil   Adenine   Guanine

Cytosine   Thymine

Drawing Nucleosides

3. Given this skeletal structure draw the nucleoside deoxythymidine.

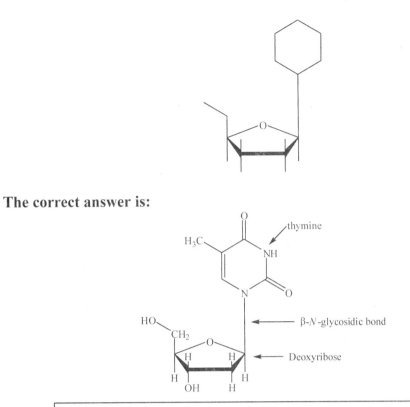

The correct answer is:

Nucleosides are made from a nucleotide bonded to Ribose or Deoxyribose via a β-*N*-glycosidic bond. Their name indicates their molecular structure. If the name begins with deoxy, the sugar is Deoxyribose. If deoxy is absent, the sugar is ribose. The latter half of the name indicates which nucleotide is bonded to the sugar - thymidine for thymine, guanosine for guanine, adenosine for adenine and so forth.

Drawing Nucleotides

4. Given this skeletal structure draw the nucleotide deoxyadenosine 5'-diphosphate.

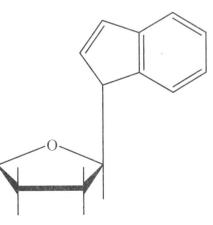

165

The correct answer is:

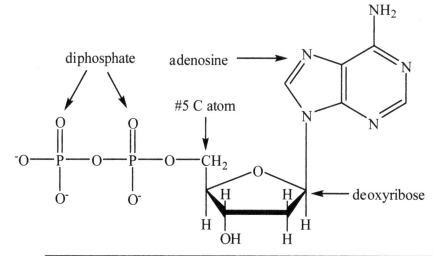

Nucleotides are composed of one of the five bases, one of the two sugars, and from one to three phosphate groups. For the nucleotide deoxyadenosine 5'-diphosphate, deoxy indicates that the sugar is Deoxyribose, 5' indicates that the phosphate groups are bonded to deoxyribose at the number 5 carbon, adenosine tells us that the base is adenine, and diphosphate describes the two phosphate groups. You must learn how to decode these names and draw the structures of nucleotides.

Primary Structure of Nucleic Acids

5. Which part of nucleotides forms the polymeric backbone of nucleic acids?
      a. heterocyclic bases   b. sugars   c. phosphate groups
   The correct answer is phosphate groups.

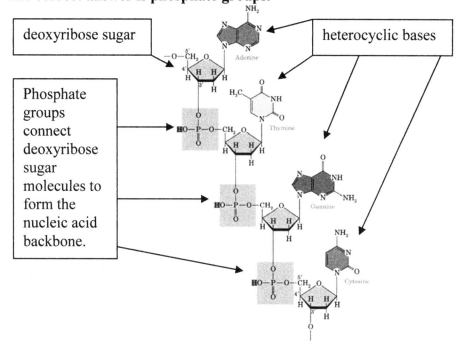

TIP | Nucleic acids are polymers composed of nucleotides. Monophosphate nucleotides bond to one another through phosphate ester linkages to form long chains. This leaves the bases exposed on one side of the chain and available to hydrogen bond with the complementary nucleic acid chain.

Hydrogen bonding in complementary base pairs

6. Shown below are the thymine and adenine complementary base pairs. Reposition the two molecules to indicate how these two molecules can maximize their hydrogen bonding interactions.

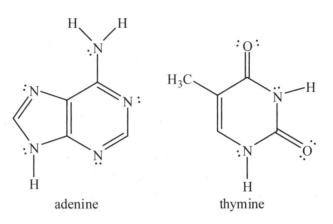

adenine   thymine

The correct answer is:

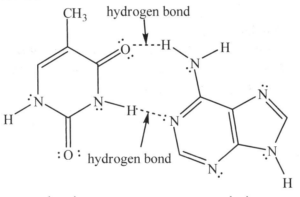

thymine   adenine

Shown below are molecular model images of the two possible complementary base pairings in DNA. This is why cytosine always pairs with guanine and adenine with thymine in the DNA double helix.

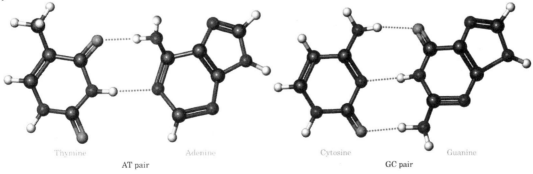

AT pair   GC pair

167

7. What is the secondary structure of DNA?
    a. solenoid structure   *b. loop structures*
    c. miniband structure   *d. double helix structure*
    e. stacked miniband structure
    The correct answer is double helix structure.

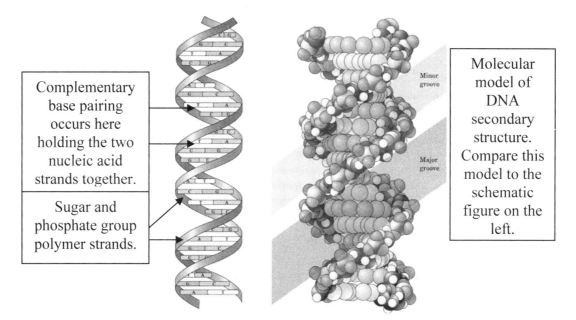

Complementary base pairing occurs here holding the two nucleic acid strands together.

Sugar and phosphate group polymer strands.

Minor groove

Major groove

Molecular model of DNA secondary structure. Compare this model to the schematic figure on the left.

Differences between DNA and RNA

8. How does RNA differ from DNA in its molecular composition and structure?

The correct answer is a) RNA replaces thymine with uracil in its complementary base pairs, b) RNA utilizes ribose instead of deoxyribose, c) RNA employs more secondary structures than DNA's double helical structure.

TIP

There are six types of RNA each of which has a specialized function in cellular processes. Consequently, their secondary molecular structures are adapted to help them accomplish their function.

Math Review

Introduction

General, organic and biochemistry classes require several basic mathematical skills. These include many which you were taught earlier in your academic career but may have forgotten from lack of use. This section will refresh your memory on the math skills necessary in the typical general chemistry course. The important topics in this section include:

1. the proper use of scientific notation
2. the rounding of numbers
3. rearranging algebraic equations
4. rules of logarithms
5. basic calculator skills, including entering numbers in scientific notation

---

Math Review Key Equations and Concepts

1. Scientific and Engineering Notation

In the physical and biological sciences it is frequently necessary to write numbers that are extremely large or small. It is not unusual for these numbers to have 20 or more digits beyond the decimal point. For the sake of simplicity and to save space when writing, a compact or shorthand method of writing these numbers must be employed. There are two possible but equivalent methods called either scientific or engineering notation. In both methods the insignificant digits that are placeholders between the decimal place and the significant figures are expressed as powers of ten. Significant digits are then multiplied by the appropriate powers of ten to give a number that is both mathematically correct and indicative of the correct number of significant figures to use in the problem. To be strictly correct, the significant figures should be between 1.000 and 9.999; however, this particular rule is frequently ignored. In fact, it must be ignored when adding numbers in scientific notation that have different powers of ten.

The only difference between scientific and engineering notation is how the powers of ten are written. Scientific notation uses the symbolism "x 10^y" whereas engineering notation uses the symbolism "Ey" or "ey". Engineering notation is frequently used in calculators and computers.

/TIPS\

Positive powers of ten indicate that the decimal place has been *moved to the left that number of spaces*.

Negative powers of ten indicate that the decimal place has been *moved to the right that number of spaces*.

A few examples of both scientific and engineering notation are given in this table.

Number	Scientific Notation	Engineering Notation
10,000	1×10^4	1E4
100	1×10^2	1E2
1	1×10^0	1E0
0.01	1×10^{-2}	1E-2
0.000001	1×10^{-6}	1E-6
23,560	2.356×10^4	2.356E4
0.0000965	9.65×10^{-5}	9.65E-5

It is important for your success in chemistry that you understand how to use both of these methods of expressing very large or small numbers. Familiarize yourself with both methods.

2. **Rounding of Numbers**

When determining the correct number of significant figures for a problem it is frequently necessary to round off an answer. Basically, if the number immediately after the last significant figure is a 4 or lower, round down. If it is a 6 or higher, round up. The confusion arrives when the determining number is a 5. If the following number is a 5 followed by a number greater than zero, round the number up. If the number after the 5 is a zero, then the textbook used in your course will have a rule based upon whether the following number is odd or even. You should use that rule to be consistent with your instructor. The following examples illustrate these ideas. In each case the final answer will contain three significant figures.

Initial Number	Rounded Number
3.67492	3.67
3.67623	3.68
3.67510	3.68
3.67502	Use your textbook rule.

3. **Rearranging Algebraic Equations**

One task that you will perform often is the rearrangement of algebraic equations to solve for a particular variable. For example, the equation for the conversion of temperature from Fahrenheit Scale to Celsius scale is

$$^{\circ}C = \frac{5}{9}(^{\circ}F - 32)$$

The equation can be rearranged to solve for °F.

$$°C = \frac{5}{9}(°F - 32)$$

multiply each side by 9
$$9 \times °C = 9 \times \frac{5}{9}(°F - 32)$$

$$9 \times °C = 5(°F - 32)$$

divide each side by 5
$$\frac{9 \times °C}{5} = \frac{5(°F - 32)}{5}$$

$$\frac{9}{5}°C = °F - 32$$

add 32 to each side
$$\frac{9}{5}°C + 32 = °F - 32 + 32$$

$$\frac{9}{5}°C + 32 = °F$$

rearrange so that °F is on the left
$$°F = \frac{9}{5}°C + 32$$

CAUTION

Be sure that you are able to rearrange equations. Having this ability will greatly simplify the number of equations you will need to memorize.

4. Logarithms
a. Definition of Logarithms

Logarithms are convenient methods of writing numbers that are exceptionally large or small and expressing functions that are exponential. They also have the convenience factor of making the multiplication and division of numbers written in scientific notation especially easy because in logarithmic form addition and subtraction of the numbers is all that is required. By definition, a logarithm is the number that the base must be raised to in order to produce the original number. There are two standard logarithms used, base 10 and base e.

General Definition of Logarithms $\quad x = a^y$ then $y = \log_a x$

Definition of Log Base 10 $\quad x = 10^y$ then $y = \log_{10} x$ or simply $\log(x)$

Definition of Log Base e $\quad x = e^y$ then $y = \log_e x$ or simply $\ln(x)$

For example, if the number we are working with is 1000 then 10, the base, must be cubed, raised to the 3rd power, to reproduce it. Mathematically, we are stating that $1000 = 10^3$, so the log (1000) = 3.

There are three commonly used rules of logarithms that you must know.

171

Multiplication Rule for Logarithms: $\log(x \cdot y) = \log(x) + \log(y)$

Division Rule for Logarithms: $\log(\frac{x}{y}) = \log(x) - \log(y)$

Exponent Rule for Logarithms: $\log(x^n) = n \log(x)$

CAUTION: While the rules shown above are in base 10, they apply equally to natural logs or any other base logs.

b. **Significant Figures for Logarithms**

There are 4 significant figures in the number 2.345×10^{12} (the 2, 3, 4, and 5). The power of 10 (the number 12) is not counted as significant. If we take the log of 2.345×10^{12} the number of significant figures must remain the same. The log of $2.345 \times 10^{12} = 12.3701$.

What numbers indicate the exponents that are present in scientific notation? In logarithms, the numbers to the left of the decimal place (the characteristic) are insignificant and the ones to the right (the mantissa) are significant. Thus, the $\log(2.345 \times 10^{12}) = 12.3701$, and both numbers have 4 significant figures.

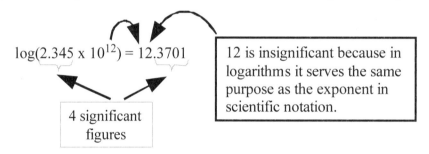

$$\log(2.345 \times 10^{12}) = 12.3701$$

4 significant figures

12 is insignificant because in logarithms it serves the same purpose as the exponent in scientific notation.

5. **Basic Calculator Skills**

General, organic, and biochemistry courses require calculations that are frequently performed on calculators. You do not need to purchase an expensive calculator for your course. Rather, you need a calculator that has some basic function keys. Common important functions to look for on a scientific calculator are: log and ln, antilogs or 10^x and e^x, ability to enter numbers in scientific or engineering notation, x^2, $1/x$, $\sqrt{}$ or multiple roots, like a cube or higher root.

More important than having an expensive calculator is knowledge of how to use your calculator. It is strongly recommended that you study the manual that comes with your calculator and learn the basic skills of entering numbers and understanding the answers that your calculator provides. For a typical general chemistry course there are three important calculator skills with which you should be proficient.

1) <u>Entering Numbers in Scientific Notation</u>
   Get your calculator and enter into it the number 2.54×10^5. The correct sequence of strokes is: press 2, press the decimal button, press 5, press 4, and *then press either EE, EX, EXP or the appropriate exponential button on your calculator. Do not press x 10 before you press the exponential button!* This is a very common mistake and will cause your answer to be 10 times too large.

CAUTION	After you have entered 2.54×10^5 into your calculator, press the Enter or = button and look at the number display. If it displays 2.54E6 or 2.54×10^6, you have mistakenly entered the number. Correct your number entering method early in the course before it becomes a bad habit!

2) <u>Taking Roots of Numbers and Entering Powers</u>
   Frequently we must take a square or cube root of a number to determine the correct answer to a problem. Most calculators have a square root button, $\sqrt{\ }$. To take a square root, simply enter the number into your calculator and press the $\sqrt{\ }$ button to get your answer. For example, take the square root of 72 (the answer is 8.49).

   Some calculators have a $\sqrt[3]{\ }$ button as well. If your calculator does not have a $\sqrt[3]{\ }$ button, then you can use the y^x button to achieve the same result. To take a cube root, enter 1/3 or 0.333 as the power and the calculator will take a cube root for you. For example, enter $27^{0.333}$ into your calculator (the correct answer is 3.00). If you need a fourth root, enter ¼ which is 0.25 as the power, and so forth for higher roots.

3) <u>Taking base 10 logs and natural or naperian logs, ln</u>
   Many of the functions in thermodynamics, equilibrium, and kinetics require the use of logarithms. All scientific calculators have log and ln buttons. To use them simply enter your number and press the button. For example, the log 1000 = 3.00, and the ln of 1000 = 6.91.

CAUTION	A common mistake is taking the ln when the log is needed and vice versa. Be careful which logarithm you are calculating for the problem.

Sample Exercises

1. *Which of the following numbers is very large or very small?*

   $3.45x10^{-9}$ $\qquad$ $3.45x10^2$ $\qquad$ $3.45x10^{-2}$ $\qquad$ $3.45x10^9$

   The correct answers are: $\qquad$ **$3.45x10^9$ is very large**
   $\qquad\qquad\qquad\qquad\qquad\qquad$ **$3.45x10^{-9}$ is very small**

   Remember that positive powers of ten means a number is larger than 1. In the case of $3.45x10^9$, this is 3.45 multiplied by 1 billion.

   Negative powers of ten means a number is smaller than 1. In the case of $3.45x10^{-9}$, this is 3.45 divided by 1 billion.

   $3.45x10^2$ in decimal notation is 34.5 and $3.45x10^{-2}$ in decimal notation is 0.0345.

2. *The ideal gas law is*

   $$PV = nRT$$

   Where P is the gas pressure, V is the volume the gas occupies, n is the number of moles of gas, R is the gas constant, and T is the temperature measured in Kelvin. Rearrange the equation to solve for the number of moles of gas, n.

   The correct answer is: $\qquad$ $n = \dfrac{PV}{RT}$

   $$PV = nRT$$

   divide each side by R $\qquad$ $\dfrac{PV}{R} = \dfrac{n\cancel{R}T}{\cancel{R}}$

   $$\dfrac{PV}{R} = nT$$

   divide each side by T $\qquad$ $\dfrac{PV}{RT} = \dfrac{n\cancel{T}}{\cancel{T}}$

   $$\dfrac{PV}{RT} = n$$

   rearrange the equation so *n* is on the left $\qquad$ $n = \dfrac{PV}{RT}$

3. Rearrange the following equation to solve for log(y).

$$\log\left(\frac{x}{y}\right) = \log(2)$$

The correct answer is:
$$\log(y) = \log\left(\frac{x}{2}\right)$$

	$\log\left(\dfrac{x}{y}\right) = \log(2)$
use the division rule of logarithms to separate x and y	$\log(x) - \log(y) = \log(2)$
subtract log(x) from each side	$\log(x) - \log(y) - \log(x) = \log(2) - \log(x)$
	$-\log(y) = \log(2) - \log(x)$
multiply both sides by -1	$-1 \times (-\log(y)) = -1 \times (\log(2) - \log(x))$
	$\log(y) = -\log(2) + \log(x)$
rearrange the left side to a normal subtraction	$\log(y) = \log(x) - \log(2)$
use the division rule to combine log(x) and log(2) into a single expression	$\log(y) = \log\left(\dfrac{x}{2}\right)$

4. Evaluate the following expression to the correct number of significant figures.

$$\log\left(5.321 \times 10^{-5}\right) = ?$$

The correct answer is: **-4.2740**

This is what the calculator provided as the answer.

$$\log(5.321 \times 10^{-5}) = -4.274006741$$

Original number has four (4) significant figures.

The numbers before the decimal are not significant when taking a logarithm.

These are the four significant figures for the logarithm.

Since the digit following the last significant digit is 0, there is no need to round.

175

5. *Perform the following calculation with your calculator.*

$$5.873 \times 10^5 + 6.23 \times 10^5 = ?$$

The correct answer is: 1.210×10^6

If your answer had a magnitude of 10^7, then you need to review the instructions for your calculator on how to enter exponents.